BEGINNER'S LEAD TO SOIL SCIENCE

Learn the Basics of Soil Composition, Health, and Management for Gardening, Agriculture, and Environmental Sustainability

By

John William Bush (The Farmer's House)

Copyright@2024 John William Bush "The Farmer's House"

All rights reserved The publisher's permission is required before this book may be copied or reproduced in any way. As a result, nothing therein can be transmitted, saved electronically, or retained in a database. The original publisher's consent is required before any part or all of the material may be copied, scanned, faxed, or stored.

TABLE OF CONTENT

Introduction 7
  Recognizing the Significance of Soil 7
    Obstacles to Soil Health 11
  An Overview of Soil Science 12
    Soil: What Is It? 13
    Functions of Soil 16

Chapter 1: Soil Structure 19
  What Constitutes Soil? 19
  Minerals' Function in Soil 25
  Organic Matter's Significance in Soil 29
    The Significance of Living Things 30
  Texture and Structure of Soil 34
    Soil Texture's Effect on Soil Properties 35
    Different Soil Structure Types 37

Chapter 2: Properties of Soil 43
  Soil pH Levels 43
    Factors Affecting the pH of Soil 45
    pH of the Soil Is Important for Agriculture 47
  Density and Porosity of Soil 51
    Calculating Soil Density 52
    Factors Influencing Density of Soil 53

Porosity of Soil 55
The Value of Porosity and Density in Soil for
Agriculture 57
Water Content of Soil 60
Definition and Importance of Soil Water Content .. 61
Calculating the Water Content of Soil 61
Soil Water Content: Its Significance in Agriculture 64
Controlling the Water Content of Soil 66

Chapter 3: Nutrients in Soil 69
Knowing macro- and micronutrients 69
Crucial Elements for the Growth of Plants 73
Overview of Soil Nutrients 73
Macronutrients That Are Required for Plant Growth
.... 74
Crucial Micronutrients for the Growth of Plants 78
Fertility of Soil and Management of Nutrients 83
Factors Influencing Fertility of Soil 85
Agriculture's Use of Nutrient Management 86

Chapter 4: Health of Soil 89
Factors Impacting the Health of Soil 89
Factors Impacting the Health of Soil 90
The Value of Healthy Soils 93
Microorganisms in Soil and Their Function 94

Soil Microorganisms' Roles 95
Elements Affecting Microorganisms in the Soil 97
Erosion of Soil and Preservation 99
Why Soil Erosion Occurs 100
Effects of Soil Degradation 101
Techniques for Preserving Soil............................. 102

Chapter 5: Examining Soils **105**
The Value of Soil Examination 105
The Advantages of Soil Analysis 107
How to Perform a Soil Test 108
How to Gather and Analyze Soil Sample Data 110
Advice on How to Sample and Interpret Soil........ 117
Utilizing Soil Test Findings for Management of Soil
.... 118
Methods of Sustainable Soil Management 124

Chapter 6: Methods of Soil Management **127**
Soil Amendments: Organic vs. Inorganic................. 127
Benefits of Adding Organic Matter to Soil 129
The drawbacks of adding organic matter to soil.... 130
Benefits of Inorganic Amendments to Soil 132
The drawbacks of adding inorganic materials to soil:
.... 132
Using Compost to Improve Soil 136

Crop rotation and cover cropping 143
   Cover Cropping 143
   Crop Rotation 146

Chapter 7: Classification and Types of Soils 149
  An Overview of Systems for Classifying Soil 149
   Types of Soil 150
   Systems for Classifying Soils 151
  The Significance of Classifying Soils 155
   Classification Criteria for Soils 156
  Principal Soil Types in the World 158
  Mapping and Interpretations of Soils 163

Chapter 8: Sustainable Soil Practices 171
  Techniques for Preserving Soil 171
  Methods of Sustainable Agriculture 176
  Management of Urban Soils 182

Conclusion: A Recap of the Main Ideas 189
  Next Developments in Soil Science 193

Glossary 198

Introduction

Recognizing the Significance of Soil

An essential element of our natural surroundings, soil is vital to the continuation of life as we know it. It serves as the basis for the health of our ecosystems by fostering the growth of plants, offering homes to a vast array of creatures, and acting as a storehouse for water and nutrients. Since soil is the foundation for both ecosystem health and human food production, it is vital knowledge for anybody interested in gardening, agriculture, or environmental sustainability.

Soil's Function in Ecosystems

Since soil makes up a large portion of the planet's land area and is essential to life support, it is frequently referred to as the "skin of the Earth." Numerous species, ranging from microscopic bacteria to larger critters like earthworms and insects, live in this intricate and ever-changing ecosystem. These creatures are essential to the health of the soil because they break down organic matter, cycle nutrients, and enhance soil structure.

Plant growth is one of the soil's primary purposes. Plants may set roots, obtain water and nutrients, and find anchorage in the soil. In addition, it acts as a medium for the exchange and storage of nutrients, giving plants access to nitrogen, phosphorus, and potassium. Plants would struggle to live in unhealthy soil, which would have a domino effect on the ecosystem as a whole.

Quality of the Environment and Soil Functions

Because it stores carbon, supports biodiversity, and filters and purifies water, soil is essential to preserving the quality of the environment. By filtering impurities and pollutants out of water as it seeps through the soil profile, soil serves as a medium. Furthermore, soil acts as a carbon sink, removing carbon dioxide from the atmosphere and reducing global warming. The diversity of life found in soil, such as microbes, insects, and earthworms, is crucial for preserving ecosystem function and soil health.

Global Cycles of Biogeochemistry

The carbon, nitrogen, and phosphorus cycles are three major worldwide biogeochemical cycles in which soil is an essential component. Large volumes of organic carbon are stored in the soil as soil organic matter, acting as a carbon reservoir. For organic matter to break down and release carbon dioxide back into the atmosphere, soil organisms are essential. Furthermore, soil plays a crucial role in the nitrogen cycle by providing a home for bacteria that fix nitrogen from the atmosphere and transform it into a form that plants can utilize. The phosphorus cycle involves soil as well since it acts as a storehouse for the mineral, which is necessary for plant growth.

Water and Soil

As a reservoir that both stores and filters water, soil is essential to the water cycle. Raindrops are absorbed by the soil, which stops flow and erosion. During dry spells, this process keeps streams flowing and replenishes groundwater sources. Particularly in arid areas, healthy soils with a high organic matter content and strong structure are better at retaining water, which is necessary for plant growth.

Climate and Soil

Additionally, soils are important in controlling the Earth's climate. By removing carbon dioxide from the atmosphere, the soil absorbs carbon dioxide and stores it as organic matter, which helps to lessen the consequences of climate change. On the other hand, inappropriate soil management techniques, such as extensive farming and deforestation, can cause soil deterioration and the release of stored carbon, which increases greenhouse gas emissions.

Security of Food and Soil

Producing food is arguably the greatest direct use of soil in human life. The vital nutrients that plants require to flourish are found in soil, and these plants subsequently produce food for both people and animals. Since healthy soil enables farmers to grow crops effectively and responsibly, it is crucial for guaranteeing food security. However, a major danger to the security of food supplies worldwide is soil degradation brought on by things like pollution, misuse, and erosion.

Obstacles to Soil Health

Despite its significance, soil is in danger of losing its health and productivity due to several issues. In many regions of the world, overgrazing, deforestation, and ineffective land management techniques are among the main causes of soil erosion. Another major worry is soil degradation brought on by the usage of agrochemicals, industrial waste, and urban runoff. In addition, salinization, loss of organic matter, and compaction of the soil are all causing the quality of the soil to deteriorate.

To sum up, soil is a valuable resource that is necessary to keep life on Earth going. It is essential for maintaining plant growth, controlling the water cycle, and sequestering carbon. But there are a lot of hazards to soil, like pollution, erosion, and deterioration, and these need to be addressed immediately. To maintain our soil's health and production for future generations, we must recognize the value of soil and implement sustainable soil management techniques.

An Overview of Soil Science

The study of soil as a natural resource, including its characteristics, development, classification, and interactions with the surrounding environment, is the focus of the multidisciplinary field of soil science. It is essential to many industries, including land management, agriculture, and environmental conservation. To ensure food security, reduce environmental degradation, and implement sustainable land use practices, one must have a solid understanding of soil science. This thorough introduction will explore every facet of soil science, from fundamental ideas to real-world applications.

Soil: What Is It?

As the Earth's surface weathers over time, a complex mixture of mineral particles, organic debris, water, air, and living things comes together to form soil. It is an active, living system that is essential to maintaining life as we know it on Earth. In addition to being a medium for plant growth, soil also acts as a filter for water, a home for a variety of species, and a repository for carbon and nutrients.

Creation of Soil

The complicated process of soil formation, or pedogenesis, is impacted by some variables, including terrain, vegetation, climate, parent material, and time. It starts with the rocks being broken down into smaller pieces by the physical and chemical weathering process. Soil is composed of these particles as well as organic matter that is derived from plant and animal carcasses. Different soil horizons or layers gradually form as a result of the interactions between these elements and the activities of soil organisms.

Properties of Soil

Numerous essential characteristics of soil affect its fertility, structure, and capacity to sustain plant growth. Texture, structure, porosity, density, pH, and nutritional content are some of these characteristics. The relative amounts of sand, silt, and clay particles in the soil determine its texture and impact its ability to retain water and nutrients. The way that soil particles are arranged into aggregates or clumps affects root penetration, water infiltration, and aeration. This process is known as soil structure. The quantity of pore space in the soil determines its porosity, which impacts its capacity to hold air and water. The mass of soil per unit volume is known as soil density, and it affects water circulation and root growth. Plant nutrient availability can be impacted by the acidity or alkalinity of the soil, which is measured by pH. The availability of vital minerals like nitrogen, phosphorus, and potassium—all of which are necessary for plant growth—is referred to as the soil's nutritional content.

Classification of Soils

The practice of grouping soils according to their attributes and features is known as soil classification. There are several categorization systems in use throughout the world, but the Soil Taxonomy system created by the US Department of Agriculture (USDA) is the most popular. Based on factors like soil texture, color, structure, and the existence of diagnostic horizons, this method divides soils into different categories. Classifying soils makes it easier to understand their characteristics and how to manage them by organizing information about them.

Functions of Soil

To support life on Earth, soil must carry out several crucial tasks. Among these roles are:

- ✓ **Providing a medium for plant growth:** Plant roots use soil as a substrate because it gives them support and anchoring. In addition, it acts as a storage space for nutrients and water, both of which are necessary for plant growth.
- ✓ **Filtering and buffering:** As water seeps through the soil profile, the soil serves as a filter, eliminating pollutants and impurities. By controlling the pH and nutrient levels of water that passes through it, soil also serves as a buffer.
- ✓ **Habitat for organisms:** A wide variety of organisms, such as bacteria, fungi, insects, and earthworms, call soil home. These creatures are

essential to the decomposition, soil formation, and cycling of nutrients.
- ✓ **Carbon storage:** Compared to all terrestrial plants and the atmosphere put together, soil is a significant carbon reservoir. Through the sequestration of carbon dioxide from the atmosphere, this carbon storage aids in the mitigation of climate change.
- ✓ **Engineering medium:** Soil serves as a sturdy foundation for roadways, buildings, and other structures when employed as an engineering medium in construction.

Degradation and Preservation of Soils

Despite its significance, the soil is under constant threat of degradation which could lower its production and quality. In many regions of the world, overgrazing, deforestation, and ineffective land management techniques are among the main causes of soil erosion. Another major worry is soil degradation brought on by the usage of agrochemicals, industrial waste, and urban runoff. In addition, salinization, loss of organic matter, and compaction of the soil are all causing the quality of the soil to deteriorate. Many soil conservation techniques, including contour plowing, terracing, cover crops, and agroforestry, have been developed to solve these problems. By using these techniques, soil erosion can be avoided, soil health can be increased, and soil resources can be preserved for future generations.

Soil science is a broad and intricate field that is essential to maintaining life as we know it on Earth. We can appreciate the role that soil plays in sustaining ecosystem health, promoting plant development, and offering vital services to society by having a basic understanding of soil science. We can save this invaluable resource for future generations by implementing sustainable land use policies and soil protection measures.

Chapter 1: Soil Structure

What Constitutes Soil?

The complex and dynamic combination of mineral particles, organic matter, air, water, and living things that make up soil. Knowing the elements that makeup soil is essential to appreciate its qualities, fertility, and capacity to foster plant growth. This chapter will examine the many elements that makeup soil and how they work together to create an environment that is both healthy and productive.

MINERAL-BASED PARTICLES

Mineral particles, which are produced as rocks and minerals weather, are among the main ingredients of soil. The three primary types of these particles are silt, clay, and sand, and their sizes vary. The largest particles, measuring between 0.05 and 2.0 millimeters, are those of sand. Large pores are created in the soil by them, which promote proper drainage and aeration. They are gritty to the touch and do not adhere to one another well. The size of silt particles ranges from 0.002 to 0.05 millimeters, making them smaller than sand. They feature a moderate water retention capacity and a pleasant feel to the touch. The smallest particles are made of clay; their size is less than 0.002 millimeters. They create compact aggregates that can obstruct water infiltration and root growth because they are sticky when wet and rigid when dry.

The relative amounts of silt, clay, and sand in a soil define its texture, which affects drainage, fertility, and the soil's ability to hold water. While sandy soils drain fast but may be deficient in nutrients, heavy clay soils tend to hold water and nutrients effectively but can also be poorly aerated.

PLANT-BASED MATERIALS

Another essential component of soil is organic matter, which is made up of partially decomposed animal and plant waste. It provides soil organisms with food and is high in nutrients. In addition to helping to retain more water in the soil, organic matter also serves to boost the availability of nutrients to plants. Additionally, because it stores carbon in the soil, it is essential for carbon sequestration, which lessens the effects of climate change.

The climate, vegetation, and land use practices are some of the elements that affect the amount of organic matter in soil. In comparison to soils in agricultural or urban settings, soils in forested areas typically contain higher levels of organic matter. Enhancing soil health and fertility can be achieved by adding organic matter to the soil through techniques including mulching, cover crops, and composting.

WATER

Water is necessary for the health of the soil and the growth of plants because it acts as a medium for microbial activity and as a solvent for nutrients. There are three types of water found in soil: hygroscopic water, which is strongly linked to soil particles and is unavailable to plants; capillary water, which is held in the soil pores against gravity; and gravitational water, which swiftly drains through the soil. A few examples of these variables are temperature, drainage, soil texture, and precipitation.

AIR

Since air provides oxygen for aerobic microbial activity and root respiration, it is also essential for the health of the soil. Soil air is a material that fills the pore spaces between soil particles. Its availability is determined by various factors, including organic matter concentration, soil compaction, and waterlogging. Sustaining microbial populations and proper root growth requires enough soil aeration.

LIVING ORGANISMS

From larger animals like earthworms and insects to microscopic bacteria and fungi, the soil is brimming with life. These creatures are essential to the decomposition, soil formation, and cycling of nutrients. Organic debris is broken down by bacteria and fungi, which release nutrients that are vital to plant growth. By tunneling through the soil and forming pathways for the passage of water and air, earthworms and other soil organisms contribute to the improvement of soil structure.

MINERAL NUTRIENTS

Soil contains nitrogen, phosphorous, potassium, and other micronutrients that are necessary for plant growth. These nutrients come from the breakdown of organic waste and the weathering of rocks and minerals. Through their roots, plants take up nutrients from the soil, and these nutrients' availability is determined by several variables, including the pH, organic matter content, and microbial activity of the soil. The ability of the soil to supply these nutrients to plants is known as soil fertility, and it is crucial for the upkeep of healthy and fruitful soils.

To sum up, soil is a dynamic and complex mixture of organic matter, water, air, and mineral particles. Every element is essential to establishing a wholesome and fruitful soil ecosystem. It is imperative to comprehend the composition of soil to effectively manage and preserve its fertility, structure, and overall health. We can guarantee the long-term sustainability of our soils for future generations by implementing methods that increase soil biodiversity, enrich soil organic matter, and improve soil structure.

Minerals' Function in Soil

Because they are vital to soil fertility, structure, and nutrient availability, minerals are an integral part of soil. Gaining knowledge about the function of minerals in soil is essential to comprehending soil health and how it affects plant development and ecosystem function. The various kinds of minerals that are present in soil, their origins, and their roles in soil systems will all be covered in this chapter.

Mineral Types in Soil

Primary minerals and secondary minerals are the two general groups into which minerals found in soil can be divided. Primary minerals are produced when rocks and other minerals weather and are generally stable over time. Primary minerals include mica, feldspar, and quartz. Primary minerals and organic materials weather both chemically and physically to produce secondary minerals. Aluminum oxides, iron oxides, and clay minerals are a few types of secondary minerals.

Minerals' Roles in the Soil

In soil systems, minerals perform many vital functions, such as:

- ✓ **Storage and Exchange of Nutrients:** Soil minerals serve as a storehouse for vital nutrients such as micronutrients, phosphorus, potassium, and nitrogen. Through a process called cation exchange, these nutrients that are stored on the surface of mineral particles can be exchanged with plant roots.
- ✓ **Soil Structure:** By aggregating or clumping together, minerals enhance the porosity, aeration, and water infiltration of the soil. Because they may form solid aggregates, clay minerals in particular are important for soil structure.
- ✓ **pH Buffering:** By acting as a buffer against pH variations in the soil, minerals in the soil can aid in keeping the growing environment for plants steady. For instance, iron and aluminum oxides can act as a buffer against alkaline conditions, while limestone, or calcium carbonate, can neutralize acidic soils.

- ✓ **Cation Exchange Capacity (CEC):** Soil's CEC, or the capacity of the soil to hold and exchange positively charged ions (cations), is influenced by minerals. Plant availability and retention of nutrients depend on CEC.
- ✓ **Water Retention:** Some minerals, like clay minerals, have a large capacity to hold water, which helps the soil hold onto moisture so that plants can use it when the weather gets dry.
- ✓ **Nutrient Availability:** By altering their solubility and mobility in the soil, minerals can have an impact on the nutrients that are available to plants. For instance, phosphate minerals can gradually release phosphorus, making it available to plants.

Sources of Soil Minerals

The weathering of rocks and minerals in the Earth's crust is the main source of minerals found in soil. Rocks are broken down into tiny pieces by weathering, which releases minerals into the soil. Because organic matter releases minerals as it decomposes, organic matter decomposition also adds to the mineral content of soil.

Mineral Content's Effects on Soil Health

The fertility and health of the soil can be significantly impacted by the mineral content of the soil. In general, mineral-rich soils are more productive and fruitful than those with low mineral content. For instance, soils with a high concentration of clay elements are often better at holding onto water and nutrients, which makes them ideal for farming. In contrast, low-mineral soils can need to have fertilizer added to them to stay fertile.

Minerals are an integral part of the soil, contributing significantly to its fertility, structure, and nutrient availability. Gaining knowledge about the function of minerals in soil is essential to comprehending soil health and how it affects plant development and ecosystem function. We can better manage and maintain soil health for sustainable agriculture and environmental protection if we acknowledge the significance of minerals in soil systems.

Organic Matter's Significance in Soil

An essential component of soil, organic matter is essential for soil fertility, health, and ecosystem function. It is rich in nutrients that are necessary for plant growth and is produced by the breakdown of plant and animal waste. We will discuss the significance of the organic matter in soil, its origins, and its effects on soil structure, nutrient cycling, and carbon sequestration.

The Significance of Living Things

In soil systems, organic matter fulfills some vital functions, including:

- ✓ **CYCLING OF NUTRIENTS:** Organic matter is a source of micronutrients, phosphorus, potassium, and nitrogen. As organic matter breaks down, these nutrients are released into the soil and become available to plants. Additionally, organic matter serves as a reservoir for nutrients, retaining them in the soil and gradually releasing them.
- ✓ **SOIL STRUCTURE:** By holding soil particles together into aggregates or clumps, organic matter contributes to a better soil structure. Plant roots will find it easier to pierce the soil and obtain water and nutrients as a result of improved soil porosity, aeration, and water infiltration.
- ✓ **WATER RETENTION:** The high water-holding capacity of organic matter aids in the soil's ability to hold onto moisture, which plants can use during dry

spells. This can increase plant tolerance to drought and lessen the requirement for irrigation.

- ✓ **ORGANIC MATTER CAN ACT AS A BUFFER AGAINST PH VARIATIONS IN THE SOIL**, preserving a steady environment that is favorable to plant growth. It can help to maintain the ideal pH levels for plant growth by neutralizing the acids and bases in the soil.
- ✓ **MICROBIAL ACTIVITY**: Earthworms, fungi, and bacteria in the soil eat organic debris as a food source. These creatures are essential to the decomposition, soil formation, and cycling of nutrients. They liberate nutrients that are vital to plant growth by dissolving organic materials into simpler molecules.
- ✓ **CARBON SEQUESTRATION:** By absorbing carbon dioxide from the atmosphere, organic matter serves as a sizable carbon sink in the soil, reducing the effects of climate change. An essential part of the global carbon cycle, soil organic matter stores

more carbon than both the atmosphere and all terrestrial vegetation put together.

Origins of Living Things

Plant and animal wastes decompose to produce organic materials in the soil. Plants deposit their leaves, roots, and other plant parts, which add to the organic matter. Animals leave behind carcasses and excrement that add to the organic materials. Additionally, microorganisms are essential to the breakdown of organic matter, reducing it to simpler molecules that are utilized by plants and other living things.

Organic Matter's Effect on Soil Health

Higher amounts of organic matter are typically linked to healthier soils, making organic matter one of the most important indicators of soil health. Rich in organic matter, soils have superior soil structure, are more fertile, and can withstand environmental stresses like erosion and drought better. In addition, a diversified soil microbial population is supported by organic matter, and soil health and nutrient cycling depend on it.

Handling of Living Things

For soil fertility and health to be maintained, organic matter management is crucial. Raising the amount of organic matter in the soil can be achieved by techniques including crop rotation, cover crops, and the addition of organic amendments like manure and compost. Additionally, by enhancing soil structure, nitrogen cycling, and water retention, these techniques can promote healthier and higher-yielding crops.

Organic matter is an essential component of soil that is critical to ecosystem function, fertility, and soil health. Comprehending the significance of organic matter in soil is imperative for sustainable land management techniques that can enhance soil fertility, augment agricultural productivity, and alleviate the effects of climate change. Acknowledging the importance of organic matter in soil systems allows us to take steps to protect and improve this priceless resource for coming generations.

Texture and Structure of Soil

Two important factors that affect the physical and chemical characteristics of soil and have an impact on its fertility, drainage, and aeration are soil texture and structure. For efficient soil management and crop production, an understanding of soil texture and structure is necessary. The principles of soil texture and structure, their measurement, and their effects on plant growth and soil health will be discussed here.

Texture of Soil

The relative amounts of sand, silt, and clay particles in a soil are referred to as its texture. The largest of these particles is sand, which is followed in size by silt and then clay. An analysis of a soil's texture reveals that its percentage of each particle size is larger in sandy soils, silty soils with higher proportions of silt, and clay soils with higher proportions of clay.

Soil Texture's Effect on Soil Properties

Many important features of soil are significantly influenced by the texture of the soil, such as:

1. WATER-HOLDING CAPACITY: Sandy soils are less able to keep water because of their greater pore spaces, which let water drain more quickly. Conversely, clay soils are better at holding water because they have fewer pore spaces. The ability of silt soils to retain water is moderate.

2. AERATION: Because of their larger pore spaces, sandy soils have better aeration than clay soils, which have smaller pore spaces. The aeration of silt soils is moderate.

3. NUTRIENT RETENTION: Clay soils are better at holding onto nutrients than sandy soils because they have a higher cation exchange capacity (CEC). Because of their reduced CEC, sandy soils might need to be fertilized more frequently.

4. WORKABILITY: Compared to clay soils, which can become heavy and sticky when wet, sandy soils are simpler to deal with and cultivate. Soils containing silt are somewhat workable.

5. EROSION RISK: Because of their loose structure and low levels of organic matter, sandy soils are more likely to erode. While clay soils are more stable, improper management can make them more prone to erosion.

Soil Organization

How soil particles are grouped into aggregates or clumps is referred to as soil structure. Microbial activity, soil texture, and organic matter content are some of the variables that affect soil structure. To encourage root development, water infiltration, and nutrient cycling, the soil structure must be healthy.

Different Soil Structure Types

Different kinds of soil structure exist, such as:

1. Granular Structure: Small, spherical aggregates that are loosely packed define granular structure. This structure is frequent in soils that have a lot of organic matter in them and is perfect for the formation of plant roots.

2. Platy Structure: Soil particles are arranged in flat, horizontal layers that define the platy structure. This structure, which is frequently present in compacted soils, can obstruct the entry of water and roots.

3. Tightly packed square or angular aggregates are the defining feature of a **blocky structure.** It is possible to find this structure in both compacted and natural soils.

4. Prismatic Structure: Soil particle pillars or columns that are vertical in orientation define a prismatic structure. This structure can obstruct root growth and is typical of soils that contain a lot of clay.

Assessing the Texture and Structure of the Soil

A soil texture triangle can be used to determine the texture of the soil. It divides the soil into different textural classes, such as loam, sandy loam, and clay, based on the percentages of sand, silt, and clay. The size, shape, and arrangement of soil aggregates in the field can be used to determine the structure of the soil, as can laboratory procedures like soil sieve analysis and soil disaggregation testing.

The Texture and Structure of Soils Are Important for Agriculture

A vital component of agriculture, soil texture, and structure affect crop growth, yield, and soil health. A farmer's decisions on crop selection, irrigation, and fertilization can be well-informed by knowing the texture and structure of the soil. For instance, sandy soils might need more regular fertilization and irrigation than clay soils, and poor-structured soils might benefit from cover crops and composting, among other methods, to enhance soil health.

1. Water Retention and Drainage: The structure and texture of the soil are very important for these processes. Sandy soils drain more quickly and have bigger pore spaces, which can cause water stress in plants during dry spells. Clay soils, on the other hand, can be more susceptible to waterlogging yet have smaller pore pores and can hold water more effectively. Water movement is also influenced by soil structure, with well-aggregated soils facilitating better drainage and infiltration of water.

2. Plant nutrient availability is influenced by the texture and structure of the soil. Because clay soils can hold onto nutrients better than sandy soils, they have a higher cation exchange capacity (CEC). Clay soils, however, can also bind nutrients, reducing their availability to plants. Because it releases nutrients during decomposition and increases nutrient availability in the soil, organic matter is essential to the cycle of nutrients.

3. Plant Health and Root Growth: Plant health and root growth are influenced by the texture and structure of the soil. Although sandy soils facilitate efficient root penetration, they may necessitate more frequent fertilizer and irrigation. Poor root development can result from clay soils' greater difficulty for roots to sift through. An ideal environment for root development and nutrient uptake is provided by well-structured soils that have a balance of sand, silt, and clay.

4. Soil Aeration: Plant roots and other soil organisms receive enough oxygen from well-aggregated soils, which is influenced by soil structure. Conversely, poor aeration in compacted soils can result in reduced microbial activity and root suffocation. The structure and aeration of the soil can be enhanced by using appropriate soil management techniques including tillage and cover crops.

5. Erosion Control: The structure and texture of the soil are essential for preventing erosion. Because of their low organic matter concentration and weak structure, sandy soils are more likely to erode. While clay soils are more stable, improper management can make them more prone to erosion. Protecting soil health and preventing erosion can be achieved by preserving soil structure through techniques like contour plowing, cover crops, and less tillage.

6. Crop Selection and Management: Crop selection and management techniques are influenced by the texture and structure of the soil. Some crops grow better in sandy soils, while others work better in clay soils. Certain crops are better suited to particular soil types. Farmers can make more informed decisions about crop selection, irrigation, fertilization, and pest control by having a thorough understanding of the texture and structure of the soil.

To sum up, soil fertility, productivity, and health are all influenced by the texture and structure of the soil. For sustainable agriculture and efficient soil management, an understanding of the texture and structure of the soil is essential. Farmers and land managers can enhance soil health and encourage sustainable land use practices by making educated decisions based on their understanding of the significance of soil texture and structure.

Chapter 2: Properties of Soil

Soil pH Levels

The basic characteristics of soil pH are vital to plant growth, nutrient availability, and soil health. For efficient soil management and sustainable agriculture, it is crucial to comprehend soil pH and how it affects various soil qualities. The idea of soil pH, how it's measured, what influences pH levels, and how important it is to agriculture will all be covered in this chapter.

pH of Soil: Definition and Importance

The concentration of hydrogen ions (H+) in the soil solution determines the pH of the soil, which indicates how acidic or alkaline the soil is. The pH scale has a neutral pH of 7, and a range of 0 to 14. Acidic soils have a pH of less than 7, whereas alkaline soils have a pH of greater than 7. A crucial element that affects a variety of soil characteristics, nutrient availability, microbial activity, and plant growth is the pH of the soil.

pH measurement of the soil

There are several ways to measure the pH of the soil, including:

1. SOIL PH METER: This instrument evaluates the electrical conductivity of the soil solution by putting a probe into the soil and taking a sample of soil. Although soil pH meters are simple to use and accurate, they need to be calibrated and maintained.

2. PH TEST STRIPS: These are paper strips that have been impregnated with dyes that are pH-sensitive and change color depending on the soil solution's pH. The pH of the soil is then ascertained by comparing the color change to a color chart. Although inexpensive and simple to use, pH test strips might not be as accurate as soil pH meters.

3. CHEMICAL INDICATORS: The pH of the soil can also be determined using chemical indicators. These indicators can be used to determine the pH of a soil sample since they change color in response to pH changes. While chemical indicators can give an approximate indication of the pH of the soil, they are not as accurate as pH meters or test strips.

Factors Affecting the pH of Soil

Soil pH can be affected by many things, including:

1. PARENT MATERIAL: Soil pH can be influenced by the kind of rock and minerals that the soil is composed of. Whereas soils formed from granite or sandstone are typically more acidic, soils originating from limestone or calcium-rich rocks are typically more alkaline.

2. CLIMATE: Through its effects on weathering rates and the decomposition of organic matter, climate can affect the pH of soil. Soils in warmer areas are typically more acidic, whilst those in drier climates are typically more alkaline.

3. ORGANIC MATTER: By affecting microbial activity and nutrient cycling, organic matter can affect the pH of soil. Organic matter decomposition can produce acids, which lowers the pH of the soil.

4. LAND USE ACTIVITIES: Soil pH can be affected by land use activities such as irrigation, liming, and fertilization. Applying lime to the soil can raise its pH, whereas ammonium-containing fertilizers can lower it.

5. MICROBIAL ACTIVITY: By producing acidic and alkaline substances, microbial activity can affect the pH of soil. Soil pH can vary due to the increased activity of specific microbial species in acidic or alkaline environments.

pH of the Soil Is Important for Agriculture

A vital component of agriculture, soil pH affects microbial activity, plant growth, and nutrient availability. The pH level of the soil has a significant impact on the nutrients that plants can get. Plants can most easily absorb most nutrients when the pH is between 6.0 and 7.5. A pH of the soil that is outside of this range might cause toxicities or nutrient shortages. The pH of the soil affects microbial activity as well. Numerous soil microorganisms are pH-sensitive and only grow in particular pH ranges. pH variations in soil can modify microbial populations and their functions, which can impact soil health and nutrient cycling. Several crucial elements of soil pH's significance in agriculture are as follows:

1. Nutrient Availability: Plants' access to vital nutrients is influenced by the pH of the soil. Soils that range from slightly acidic to neutral (pH 6.0–7.0) provide plants with the greatest availability of most plant nutrients, including micronutrients, phosphorus, potassium, and nitrogen. The availability of nutrients may be restricted beyond this range. For instance, micronutrients such as iron, manganese, and zinc may become less available in acidic soils, while phosphorus may become less available in alkaline soils. To guarantee that plants have access to the nutrients they require for healthy growth and development, the right pH range of the soil must be maintained.

2. Microbial Activity: The pH of the soil affects the activity of microorganisms in the soil, which are essential to the breakdown of organic matter and the cycling of nutrients. Numerous kinds of soil bacteria and fungi are sensitive to pH, with some thriving in a range of pH values. Sustaining soil fertility and general soil health depends on a varied and active microbial community, which can be fostered by keeping the pH level at an appropriate level.

3. Aluminum Toxicity: Aluminum can become more soluble and harmful to plants in acidic soils (pH less than 5.5). Excessive aluminum concentrations can harm plant roots and prevent them from absorbing nutrients, which results in stunted development and lower crop production. By increasing the pH and decreasing the solubility of aluminum, liming acidic soils can aid in lowering the toxicity of aluminum.

4. Soil Structure: The stability of soil aggregates is impacted by soil pH, which in turn affects soil structure. Clay particles disperse and soil aggregates break down in acidic soils due to the increased solubility of iron and aluminum oxides. This may lead to increased erosion, less water infiltration, and poor soil structure. Soil health can be enhanced and soil structure preserved by keeping the pH level at a suitable level.

5. Crop Selection: The kinds of crops that can be effectively cultivated in a specific place might be influenced by the pH of the soil. While certain crops favor somewhat alkaline environments, others are more tolerant of acidic soils. Farmers may choose the best crops for their soil conditions and maximize crop yields by being aware of the pH requirements of various crops.

6. Fertilizer Efficiency: The effectiveness of fertilizers applied to the soil can be impacted by the pH of the soil. Certain nutrients, including potassium and phosphorus, may be less accessible to plants in acidic soils, necessitating larger fertilizer application rates to achieve the same degree of nutrient uptake. Achieving the ideal pH range for the soil can enhance fertilizer efficacy and lower the chance of nutrient leaching.

To sum up, soil pH plays a crucial role in crop selection, microbial activity, soil structure, and nutrient availability in agriculture. Maintaining soil fertility, maximizing crop yields, and advancing sustainable agricultural techniques all depend on tracking and controlling the pH of the soil.

Controlling the pH of the Soil

Sustaining soil fertility and health requires careful pH management. Acidic soils can benefit from lime application to increase pH and enhance nutrient availability. For alkaline soils, sulfur or acidifying fertilizers can be applied to reduce pH. To track soil pH and identify when pH modifications are required, routine soil testing is crucial.

Microbial activity, plant growth, soil health, and nutrient availability are all significantly impacted by the pH of the soil. For efficient soil management and sustainable agriculture, it is crucial to comprehend soil pH and how it affects various soil qualities. Farmers may increase crop yields, support microbial communities that are in good health, and maximize nutrient availability by monitoring and controlling the pH of their soil.

Density and Porosity of Soil

Two crucial characteristics that affect soil health, water retention, root development, and nutrient availability are soil density and porosity. For sustainable agriculture and efficient soil management, it is essential to comprehend these characteristics. Now we will examine the ideas of soil porosity and density, as well as how they are measured, what influences them, and how they affect the quality of the soil.

Density of Soil

The mass of soil per unit volume is known as soil density, and it is commonly measured in kilograms per cubic meter (kg/m3) or grams per cubic centimeter (g/cm3). Compaction, moisture content, soil texture, and organic matter content are some of the variables that affect soil density. Density values are higher in compacted soils and lower in well-aerated soils.

Calculating Soil Density

There are several ways to estimate soil density, including:

1. CORE SAMPLING: Using a soil corer, soil cores are taken out of the earth for core sampling. Bulk density, or the total amount of soil per unit volume including both solid and pore space, is then calculated by weighing and measuring the soil cores.

2. SAND CONE METHOD: This method is employed to ascertain the soil density in the field. The volume of a hole that has been excavated is measured. The amount of dry sand supplied is then measured after a known volume of sand is put into the hole. The volume of the hole is determined by comparing the volumes before and after adding sand. The mass of the soil is divided by the hole's volume to get the density of the soil.

3. WATER DISPLACEMENT METHOD: This technique entails measuring the capacity of a water-filled container before filling it with a specified quantity of soil. The volume of the soil is determined by the increase in the volume of water that the soil displaces. One can determine the density of soil by dividing its mass by volume.

Factors Influencing Density of Soil

Soil density can be affected by many factors, including:

1. SOIL TEXTURE: Sandier soils tend to have lower density values than clayey soils, and soil texture has an impact on soil density. This is because clayey soils have more compaction and narrower pore spaces, which result in higher density values.

2. CONTENT OF ORGANIC MATTER: By enhancing soil structure and raising soil porosity, organic matter can lower soil density. By acting as a binding agent, organic matter holds soil particles in aggregates and creates pore holes that allow water and air to pass through.

3. COMPACTION: By decreasing pore gaps and enhancing soil particle packing, soil compaction can raise soil density. Foot traffic, cattle, and heavy machinery can all contribute to compaction, which reduces soil porosity and hinders root growth.

4. MOISTURE CONTENT: Wet soils often have lower density values than dry soils, therefore soil moisture content can have an impact on soil density. This is because water increases the porosity of soil by filling the pore spaces between soil particles, which decreases soil particle packing.

Porosity of Soil

The volume of pore spaces in soil is referred to as soil porosity, and it is represented as a percentage of the total volume of soil. Various elements, including soil texture, structure, compaction, and organic matter content, have an impact on soil porosity. For soil to retain water, establish roots, and support microbial activity, it must be porous.

Calculating Soil Porosity

There are several ways to measure the porosity of soil, including:

1. Bulk Density and Particle Density: As previously mentioned, core samples are used to determine bulk density. The density of the mineral particles in the soil, minus the pore spaces, is known as particle density. The difference between bulk and particle densities is divided by particle densities to determine porosity, which is then given as a percentage.

2. Water Content Method: This technique entails soaking a sample of soil and then calculating how much water the soil can hold. To calculate the volume of pore spaces in a soil sample, the volume of water retained is deducted from the sample's overall volume. The volume of pore spaces divided by the total volume of the soil sample is then used to compute porosity, which is then represented as a percentage.

Elements That Impact Soil Porosity

Soil porosity can be affected by several things, such as:

1. Soil Texture: The porosity of soil is influenced by its texture; sandy soils often have higher porosity ratings than clayey soils. This is so that more water can penetrate and roots can grow since sandy soils have bigger pore spaces and less compaction.

2. Content of Organic Matter: By strengthening the soil's structure and forming stable aggregates, organic matter can raise the porosity of the soil. Additionally, the presence of organic matter promotes microbial activity, which increases the soil's porosity by generating macropores.

3. Compaction: By constricting pore spaces and lowering water and air circulation, soil compaction lowers soil porosity. Foot traffic, cattle, and heavy machinery can all contribute to compaction, which reduces soil porosity and hinders root growth.

4. Soil Structure: Well-aggregated soils have higher porosity values than poorly aggregated soils. Soil structure affects soil porosity—stable aggregates in well-aggregated soils form macropores, which facilitate increased root development and water infiltration.

The Value of Porosity and Density in Soil for Agriculture

In agriculture, soil density and porosity are important factors that affect root development, water retention, nutrient availability, and general soil health. Here are some salient features of their significance:

1. Water Retention: The porosity of the soil influences its capacity to hold onto water. Because they can retain more water, soils with higher porosity values help plants be less vulnerable to drought stress. Sufficient soil moisture levels for plant growth depend on proper soil porosity.

2. Root Development: Soil compaction and porosity have an impact on root development through

Aeration: Low porosity and compacted soils inhibit root growth and penetration, which results in poor nutrient uptake and lower crop yields. Proper porosity and well-aerated soils encourage strong root development and nutrient uptake.

3. Nutrient Availability: By affecting the flow of water and nutrients through the soil, soil porosity influences nutrient availability. Plant roots can move and absorb nutrients more readily in soils with a high porosity. For plants to have access to the nutrients they require for proper growth and development, the density and porosity of the soil must be balanced.

4. Microbial Activity: Because it gives soil organisms a place to live and oxygen to breathe, soil porosity affects microbial activity. Microorganisms are essential for the breakdown of organic matter and the cycling of nutrients, which improves soil fertility and general health. Healthy microbial populations and nutrient cycling in the soil depend on proper soil porosity.

5. Soil Compaction: Plant development and soil health may be negatively impacted by soil compaction, which is influenced by soil density. Reduced porosity and aeration in compacted soils result in poor nutrient uptake and root development. Reduced tillage and cover crops are two examples of good soil management techniques that can help keep soil healthy and minimize compaction.

Soil porosity and density are important characteristics that affect soil health, root development, water retention, and nutrient availability in agriculture. Comprehending these attributes and their influence on the quality of soil is crucial for efficient soil management and sustainable agricultural output. Farmers may enhance crop yields, support sustainable agricultural practices, and optimize soil health by tracking and regulating soil density and porosity.

Water Content of Soil

One important factor affecting plant development, soil health, and agricultural output is the amount of water in the soil. Sustainable agriculture and efficient soil management depend on an understanding of soil water content and how to manage it. The idea of soil water content, its measurement, influencing factors, and its significance in agriculture will all be covered here.

Definition and Importance of Soil Water Content

The amount of water in the soil is referred to as its water content, and it is typically stated as a percentage of the soil's overall weight. The dynamic property of soil water content is subject to variation over time as a result of various causes, including plant absorption, evaporation, and precipitation. Because soil water content influences root development, nutrient availability, and soil structure, it is essential for plant growth.

Calculating the Water Content of Soil

There are several ways to measure the water content in soil, including:

1. Gravimetric Method: This technique takes soil samples, weighs them to find their dry weight, and then dries them out in an oven to eliminate any remaining moisture. The water content of the soil sample is determined by dividing its dry weight by its moist weight.

2. Tensiometers: Tensiometers are tools used to quantify the amount of water retained in soil by measuring the soil water potential. Tensiometers are made out of a pressure gauge coupled to a porous ceramic cup. The tension in the soil water is measured by the pressure gauge, and this measurement can be used to determine the soil water content.

3. Time Domain Reflectometry (TDR): TDR measures the soil's dielectric constant, which is correlated with the water content of the soil, using electromagnetic waves. Soil water content is determined by implanting TDR sensors into the soil and timing the electromagnetic wave's passage through the soil.

The Soil Water Content Factors

Soil water content can be influenced by several things, such as:

1. Soil Texture: The water content of the soil is influenced by its texture, with sandy soils often holding less water than clayey soils. This is because clayey soils are better at holding onto water due to their larger surface area and smaller pore gaps.

2. Organic Matter Content: By enhancing soil structure and raising soil porosity, organic matter can raise soil water content. Like a sponge, organic matter absorbs water and releases it gradually into the roots of plants.

3. Soil Structure: By affecting soil porosity and pore size distribution, soil structure influences soil water content. Larger pore pores in well-aggregated soils provide better water infiltration and retention.

4. Climate: The impact of climate on precipitation and evaporation rates affects the amount of water in the soil. Soil water content is often higher in wet climates than in dry ones.

5. Land Use Practices: Crop choice, tillage, and irrigation are examples of land use techniques that can alter the water content of the soil. While tillage can lower soil water content by increasing soil compaction and breaking up soil aggregates, irrigation can raise soil water content.

Soil Water Content: Its Significance in Agriculture

In agriculture, soil water content plays a crucial role in determining plant growth, nutrient availability, and soil health. Here are some salient features of its significance:

1. PLANT GROWTH: Because soil water content influences photosynthesis, root growth, and nutrient uptake, it is critical for plant growth. To guarantee that plants have access to the water they require for healthy growth and development, the soil's water content must be adequate.

2. NUTRIENT AVAILABILITY: Plants' access to nutrients is influenced by the water content of the soil. Plant roots absorb dissolved nutrients from soil water. To guarantee that nutrients are available to plants in the appropriate amounts and at the appropriate times, proper soil water content is crucial.

3. SOIL HEALTH: Microbial activity and soil structure are impacted by soil water content. Microbial activity is facilitated by a proper soil water content and is necessary for the breakdown of organic matter and the cycling of nutrients. The amount of water in the soil also influences the structure of the soil; well-aerated soils have better drainage and structure.

4. CROP OUTPUT: A key determinant of crop output is the water content of the soil. High yields and the avoidance of crop stress during dry spells depend on the soil's water content. To maximize crop production, effective irrigation and soil water management techniques are crucial.

5. EROSION CONTROL: By altering the stability and structure of the soil, soil water content influences soil erosion. Because they have superior structure and are less likely to be washed away by rain, soils with the right amount of water content are less likely to erode.

Controlling the Water Content of Soil

Sustaining soil health and maximizing crop yields require careful management of the amount of water in the soil. Here are several methods for controlling the amount of water in the soil:

1. IRRIGATION: During dry spells, irrigation can be utilized to augment the water content of the soil. Water application techniques and scheduling must be done correctly to guarantee effective and efficient water distribution.

2. MULCHING: By lowering evaporation and preserving soil moisture levels, mulching can aid in the conservation of soil water content. Compost or straw are examples of organic mulches that can enhance soil structure and promote water infiltration.

3. COVER CROPPING: By enhancing soil structure and adding organic matter, cover crops can help to increase the water content of the soil. Additionally, cover crops enhance the health of the soil and lessen erosion.

4. SOIL MANAGEMENT TECHNIQUES: By preserving soil structure and lowering compaction, soil management techniques including crop rotation and less tillage can help to increase soil water content. These methods support nitrogen cycling and soil health as well.

To sum up, soil water content is an important factor that affects plant development, soil health, and agricultural productivity. Sustainable agriculture and efficient soil management depend on an understanding of soil water content and how to manage it. By tracking and controlling the amount of water in the soil, farmers can maximize plant growth, strengthen the health of the soil, and increase agricultural yields.

Chapter 3: Nutrients in Soil

Knowing macro- and micronutrients

Plants need two types of vital nutrients for growth and development: macronutrients and micronutrients. These nutrients, which are derived from the soil via the plant's root system, are essential to many physiological functions in plants.

MACRONUTRIENTS:

Definition: Essential elements plants need in relatively high amounts for growth and development are known as macronutrients.

Types: Nitrogen (N), phosphorus (P), potassium (K), calcium (Ca), magnesium (Mg), and sulfur (S) are the six macronutrients.

Features:

- **Nitrogen (N):** Necessary for the production of chlorophyll, protein synthesis, and general plant growth.
- **Phosphorus (P):** Essential for photosynthesis, root development, and energy transfer.
- **Potassium (K):** Required for disease resistance, water control, and enzyme activation.
- **Calcium (Ca):** Necessary for the synthesis of cell walls, cell division, and absorption of nutrients.
- **Magnesium (Mg):** Required for photosynthesis, enzyme activation, and the production of chlorophyll.
- **Sulfur (S):** Essential for enzyme activation, protein synthesis, and general plant growth.

Sources: Plants take up macronutrients from the soil through their roots, where they are mainly taken up as ions.

Deficiency Symptoms: Growth retardation, leaf yellowing, decreased yield, and general poor plant health can result from deficiencies in micronutrients.

MICRONUTRIENTS:

Definition: Micronutrients are necessary substances that plants need in lower amounts to grow and develop.

Kinds: Iron (Fe), manganese (Mn), zinc (Zn), copper (Cu), boron (B), molybdenum (Mo), chlorine (Cl), nickel (Ni), and cobalt (Co) are the nine micronutrient types (although cobalt is not thought to be necessary for all plants).

Features:

- **Iron (Fe):** Required for the synthesis of chlorophyll and the activation of enzymes.
- **Manganese (Mn):** Required for nitrogen metabolism, enzyme activation, and photosynthesis.
- **Zinc (Zn):** Essential for the production of proteins, growth control, and enzyme activation.
- **Copper (Cu):** Needed for the construction of cell walls, electron transport, and enzyme activation.
- **Boron (B):** Necessary for sugar transport, pollen germination, and cell wall production.

- **Molybdenum (Mo):** Required for the activation of enzymes and the fixation of nitrogen.
- **Chlorine (Cl):** Essential for osmotic control and photosynthesis.
- **Nickel (Ni):** Necessary for some plants' metabolism of nitrogen.
- **Cobalt (Co):** In leguminous plants, Cobalt helps fix nitrogen.

Sources: Micronutrient deficits may arise if the soil is deficient in certain nutrients, which plants typically only receive in trace amounts from the soil.

Inadequacy Signs: Depending on the micronutrient, deficiencies can cause a variety of symptoms, such as diminished growth, necrosis, and chlorosis.

Sufficient concentrations of macronutrients and micronutrients must exist in the soil for plants to develop and survive. For sustainable agriculture and efficient soil management, it is crucial to comprehend the functions and needs of these nutrients.

Crucial Elements for the Growth of Plants

Because soil nutrients supply the building blocks for plant development, metabolism, and reproduction, soil nutrients are critical to plant growth. For sustainable agriculture and efficient soil management, it is imperative to comprehend the function of vital nutrients in the soil and how they affect plant growth. The idea of soil nutrients, as well as the nutrients that are necessary for plant growth, their sources, purposes, and symptoms of deficiencies, will all be covered here.

Overview of Soil Nutrients

Chemical components known as soil nutrients are necessary for the growth and development of plants. The two groups of nutrients that they belong to are macronutrients and micronutrients. Plants need macronutrients more than micronutrients, which are needed in lesser proportions. Numerous plant functions, such as photosynthesis, respiration, and nutrient uptake, depend heavily on soil nutrients.

Macronutrients That Are Required for Plant Growth

1. NITROGEN (N):

- ➤ *Function:* A vital component of proteins, amino acids, and chlorophyll, nitrogen is necessary for plant growth. It is essential for respiration, photosynthesis, and nucleic acid production.
- ➤ *Sources:* Plants mainly take up nitrogen from the soil as nitrate (NO_3^-) or ammonium (NH_4^+) ions. Soil nitrogen is also influenced by organic matter and bacteria that fix nitrogen found in legume root nodules.
- ➤ *Deficiency Symptoms:* When plants lack nitrogen, their growth is inhibited, their leaves turn yellow (a condition known as chlorosis), and their production is decreased.

2. PHOSPHORUS (P):

- *Function:* Phosphorus is a component of phospholipids, nucleic acids, and ATP (adenosine triphosphate), and it is necessary for plant growth. It is essential for cell division and the transfer of energy.
- *Sources:* Plants primarily take up phosphorus from the soil as phosphate (PO_4^{3-}) ions. Organic matter and phosphate minerals are significant sources of phosphorus in soil.
- *Shortage Symptoms:* Plants with a phosphorus shortage grow slowly, bloom later, and have poor root development.

3. KALIUM (K):

- *Function:* Potassium is essential for protein synthesis, enzyme activation, and plant water uptake. It also plays a role in preserving osmotic equilibrium and cell turgor pressure.

- **Sources:** Plants typically take up potassium from the soil as potassium ions (K^+). Soil potassium is derived from organic matter and potassium minerals.
- **Inadequacy Signs:** Plants lacking in potassium experience wilting, decreased fruit quality, and yellowing of the leaf edges.

4. Calcium (Ca):

- **Function:** Enzyme activation, cell wall synthesis, and membrane stability all depend on calcium. It also contributes to nutrient absorption and plant signaling.
- **Sources:** Plants primarily take up calcium from the soil as calcium ions or Ca^{2+}. Soil calcium is influenced by the minerals calcium phosphate and carbonate.
- **Deficiency Symptoms:** Plants experiencing a calcium deficiency exhibit stunted growth, deformed leaves, and fruit blossom end rot.

5. MAGNESIUM:

- *Function:* Magnesium is necessary for photosynthesis and is a part of chlorophyll. It also contributes to the production of nucleic acids and the activation of enzymes.
- *Sources:* Plants mostly absorb magnesium from the soil in the form of magnesium ions (Mg^{2+}). Soil magnesium is influenced by minerals that contain magnesium.
- *Deficit Symptoms:* Older leaves that have chlorosis, stunted growth, and poor fruit development are signs of a magnesium deficit in plants.

6. SULFUR (S):

- *Function:* Proteins, coenzymes, and amino acids all contain sulfur. It is involved in photosynthesis and nitrogen metabolism and is necessary for plant growth.

- ➤ *Sources:* Plants mostly take up sulfur from the soil as sulfate (SO_4^{2-}) ions. Soil sulfur is caused by sulfur-containing minerals and organic materials.
- ➤ *Deficiency Symptoms:* Yellowing of the leaves, slowed growth, and decreased yield are the symptoms of a sulfur deficiency in plants.

Crucial Micronutrients for the Growth of Plants

1. FERROUS IRON (FE):

- ➤ *Function:* Enzyme activation, photosynthesis, and chlorophyll synthesis all depend on iron. It is essential to the transfer of electrons and the creation of energy.

- ➢ *Sources:* Plants mostly take up iron from the soil as ferrous (Fe^{2+}) or ferric (Fe^{3+}) ions. Soil iron is derived from organic materials and iron oxides.
- ➢ *Shortage Symptoms:* Young leaves that exhibit chlorosis, stunted growth, and yellowing of the interveinal tissue are signs of an iron shortage in plants.

2. THE MANGANESE (MN) LANGUAGE

- ➢ *Function:* Enzyme activity, nitrogen metabolism, and photosynthesis all depend on manganese. It is essential for the production of antioxidants and chlorophyll.
- ➢ *Sources:* Plants mostly take up manganese from the soil as manganese ions (Mn^{2+}). Soil manganese is a result of both organic matter and manganese minerals.
- ➢ *Deficiency Symptoms:* Plants with a manganese deficiency exhibit interveinal chlorosis, stunted growth, and inadequate root formation.

3. ZINC (ZN):

- *Function:* Hormone control, protein synthesis, and enzyme activation all depend on zinc. It is essential to the growth and development of plants.
- *Sources:* Plants mostly absorb zinc from the soil in the form of zinc ions (Zn^{2+}). Soil zinc is derived from organic matter and zinc minerals.
- *Deficit Symptoms:* Plants experiencing a zinc deficit exhibit interveinal chlorosis, slowed development, and decreased yield.

4. COPPER (CU):

- *Function:* Enzyme activity, photosynthesis, and cell wall construction all depend on copper. It is essential to the growth and reproduction of plants.
- *Sources:* Plants mostly take up copper from the soil as copper ions (Cu^{2+}). Soil copper is derived from organic matter and copper minerals.

> ➤ *Deficit symptoms:* Young leaves with chlorosis, wilting, and decreased growth are signs of a copper deficit in plants.

5. BORON (B):

> ➤ *Function:* Hormone modulation, glucose metabolism, and the production of cell walls all depend on boron. It is essential to the growth and development of plants.
>
> ➤ *Sources:* Borate (BO_3^{3-}) ions from the soil are the main way that plants absorb boron. Soil boron is a result of both organic matter and boron minerals.
>
> ➤ *Deficiency Symptoms:* When plants are lacking in boron, they exhibit twisted leaves, slowed growth, and less fruit formation.

6. MAGNESIUM (MO):

- ➤ **Function:** Sulfur metabolism, enzyme activation, and nitrogen fixation all depend on molybdenum. It is essential to the nitrogen cycle and plant growth.
- ➤ **Sources:** Plants mostly absorb molybdenum from the soil as molybdate (MoO_4^{2-}) ions. Organic materials and minerals containing molybdenum are the sources of soil molybdenum.
- ➤ **Deficiency Symptoms:** Plants lacking molybdenum exhibit stunted growth, poor nitrogen metabolism, and yellowing of the older leaves.

Nutrients in the soil are vital for plant growth and are important for many plant functions. Effective soil management and sustainable agriculture depend on an understanding of the functions of critical nutrients in the soil and how they affect plant growth. Farmers can maximize plant growth, improve soil health, and increase agricultural productivity by making sure plants have access to the nutrients they need.

Fertility of Soil and Management of Nutrients

The ability of the soil to support plant development and maintain agricultural production is determined by its fertility, making it a critical component of agriculture. Numerous elements, such as soil nutrients, organic matter concentration, pH, and soil structure, affect soil fertility. To preserve soil fertility and maximize crop productivity, soil nutrients must be managed effectively. The idea of soil fertility, the contribution of soil nutrients to soil fertility, and techniques for managing soil nutrients in agriculture will all be covered in this chapter.

Knowing the Fertility of Soil

The ability of the soil to supply necessary nutrients to plants in sufficient amounts and ratios for healthy growth and development is referred to as soil fertility. Both natural and man-made elements, such as soil parent material, climate, topography, land use, and soil management techniques, have an impact on soil fertility. Because it guarantees the long-term production of agricultural fields, maintaining soil fertility is crucial for sustainable agriculture.

The Soil Fertility Role of Nutrients

Since they are necessary for plant growth and development, soil nutrients are an important component of soil fertility. There are two types of nutrients: macronutrients and micronutrients. While micronutrients like iron, manganese, and zinc are needed in lesser amounts, macronutrients like nitrogen, phosphorous, and potassium are needed in larger numbers. Plants absorb nutrients from the soil through their root systems, and these nutrients are necessary for photosynthesis, respiration, and nutrient uptake, among other plant functions.

Factors Influencing Fertility of Soil

Soil fertility can be impacted by some factors, including:

1. Soil Nutrients: One important factor influencing soil fertility is the presence of vital nutrients in the soil. Sufficient amounts of nutrients are necessary for the best possible growth and development of plants.

2. Organic Matter Content: As it gives plants nutrition, strengthens soil structure, and increases water retention, organic matter is an essential part of soil fertility.

3. pH Level: The availability of nutrients is influenced by the pH of the soil, with slightly acidic to neutral soils (pH 6.0–7.0) having higher nutrient availability. An extreme pH can restrict the availability of nutrients and have an impact on plant growth.

4. Soil Structure: Water infiltration, root penetration, and nutrient uptake are all impacted by soil structure, which in turn affects soil fertility. Soils with good structure and good aggregation are more fertile than those with poor structure.

5. Microbial Activity: To maintain soil fertility, soil microorganisms are essential to the cycling of nutrients and the breakdown of organic matter. To keep soil fertile, healthy microbial populations are necessary.

Agriculture's Use of Nutrient Management

To preserve soil fertility and maximize crop productivity, soil nutrients must be managed effectively. The goal of nutrient management strategies is to guarantee that plants receive the right nutrients at the right times and in the right amounts. In agriculture, many nutrient management techniques are employed, such as:

1. Soil Testing: Since it tells us about the nutritional condition of the soil, soil testing is an essential part of nutrient management. Farmers can use soil tests to assist them choose the best nutrient management strategies for their crops.

2. Application of Fertilizer: Fertilizers are applied to enhance soil fertility and supplement soil nutrients. There are several varieties of fertilizers available, including organic and inorganic fertilizers, each with a unique nutrient composition and application needs.

3. Crop Rotation: This technique involves growing a variety of crops back-to-back on the same piece of land. Crop rotation reduces the accumulation of pests and diseases and diversifies nutrient demands, all of which contribute to increased soil fertility.

4. Cover Cropping: When the primary crop isn't growing, cover crops like clovers, grasses, or legumes are sown in their place. By fixing nitrogen and introducing organic matter to the soil, cover crops aid in enhancing soil fertility.

5. Application of Manure and Compost: Rich in nutrients, organic resources like manure and compost can raise soil fertility. To enhance soil structure and supplement soil nutrients, they are frequently added to soil.

6. Precision Agriculture: To improve nutrient management techniques, precision agriculture makes use of technology like GPS and remote sensing. This minimizes nutrient losses and enables farmers to apply fertilizers more effectively.

Soil fertility is critical to sustainable agriculture since it establishes the soil's capacity to maintain agricultural production and encourage plant growth. Maintaining soil fertility and maximizing crop productivity depend on the efficient management of soil nutrients, which are a major factor in soil fertility. Farmers may improve soil health and agricultural sustainability by employing nutrient management strategies to guarantee that their crops have access to the nutrients they need for optimal growth and development.

Chapter 4: Health of Soil

Factors Impacting the Health of Soil

A vital component of agriculture is soil health, which establishes the soil's capacity to support biodiversity, encourage plant development, and preserve ecosystem services. Land management techniques and the physical, chemical, and biological characteristics of the soil all have an impact on soil health. The idea of soil health, variables influencing soil health, and the significance of preserving healthy soils for sustainable agriculture will all be covered in this chapter.

Knowing the Health of the Soil

The ability of soil to function as a living ecosystem that supports people, animals, and plants is referred to as soil health. Elevated amounts of organic matter, a wide range of microbial communities, a well-organized soil structure, and an ideal nutrient cycle are attributes of healthy soils. Because healthy soil promotes plant growth, increases nutrient and water retention, and strengthens ecosystem resilience, it is crucial for sustainable agriculture.

Factors Impacting the Health of Soil

Soil health can be impacted by many factors, including:

1. Soil Organic Matter: Because it gives plants nutrients, strengthens the soil's structure, and increases water retention, soil organic matter is a major factor in determining the health of the soil. Soil bacteria break down organic matter, which comes from leftover plant and animal debris, to release nutrients.

2. Soil pH: A measure of the acidity or alkalinity of the soil, soil pH can impact microbial activity, soil structure, and nutrient availability, all of which are factors in soil health. For best growth, most plants prefer slightly acidic (pH 6.0–7.0) soils over neutral ones.

3. Soil Structure: The way soil particles are arranged into clumps or aggregates is referred to as soil structure. Because it facilitates improved nutrient exchange, root penetration, and water infiltration, a healthy soil structure is essential to soil health. Land management techniques, soil texture, and organic matter concentration are a few examples of the variables that might affect soil structure.

4. Soil Texture: The proportions of sand, silt, and clay particles in the soil are referred to as the texture of the soil. Because soil texture influences drainage, nutrient availability, and water retention, it has an impact on soil health. While clay soils contain smaller particles and retain water longer, sandy soils tend to have larger particles and drain more quickly.

5. Soil Compaction: When soil particles are compressed, pore space is reduced, which inhibits root development and nutrient uptake. Excessive tillage, livestock grazing, and heavy machinery can all contribute to soil compaction, which can be detrimental to the health of the soil.

6. Soil Erosion: The loss or displacement of soil due to wind, water, or other causes is known as soil erosion. The removal of topsoil, which is rich in nutrients and organic matter, can result in soil erosion and a decline in soil health. Poor land management techniques, such as overgrazing and deforestation, can make soil erosion worse.

7. Soil Contamination: When dangerous materials like pesticides, heavy metals, and industrial pollutants build up in the soil, it can lead to contamination. Because contaminated soils can be dangerous to people, animals, and plants, they can have a detrimental effect on both the health of the soil and human health.

8. Land Management Techniques: Techniques including tillage, crop rotation, cover crops, and irrigation can have an impact on the quality of the soil. Organic farming and decreased tillage are two sustainable land management techniques that can enhance soil health and agricultural sustainability.

The Value of Healthy Soils

Ecosystem health and sustainable agriculture depend on maintaining the health of the soil. Plant development is facilitated by healthy soils, which also increase biodiversity and improve nutrient and water retention. Farmers may increase crop yields, decrease their reliance on artificial inputs, and advance long-term sustainability by preserving the health of their soil.

The health of the soil is essential to both agriculture and ecosystems. Numerous elements, such as soil organic content, pH, structure, texture, compaction, erosion, contamination, and land management techniques, all have an impact on soil health. Farmers may improve soil health, increase agricultural production, and foster ecosystem resilience by comprehending the elements affecting soil health and putting sustainable land management strategies into practice.

Microorganisms in Soil and Their Function

Due to their various roles in promoting soil fertility, plant growth, and ecosystem sustainability, soil microbes are essential to soil health. The varied realm of soil microorganisms, their functions in soil health, and the variables affecting their activity and abundance will all be discussed.

Overview of Microorganisms in Soil

Microscopic creatures that reside in the soil and are essential to its health are known as soil microorganisms. These microorganisms include, among others, nematodes, bacteria, fungi, and archaea. Numerous functions, including the cycling of nutrients, the breakdown of organic matter, and the prevention of illness, are facilitated by soil microbes. They are necessary to sustain plant growth and preserve soil fertility.

Soil Microorganisms' Roles

1. Nutrient Cycling: By decomposing organic matter and releasing nutrients into the soil, soil microbes are essential to the process of nutrient cycling. They break down organic stuff, including decaying plant and animal remains, and transform it into forms that are useful to plants for growth.

2. Decomposition of Organic Matter: Animal dung and plant leftovers are among the organic materials in the soil that soil microbes break down. Nutrients are released back into the soil during this decomposition process, where plants can absorb them.

3. Nitrogen Fixation: Several soil microorganisms, including bacteria and archaea that fix nitrogen from the atmosphere, can transform atmospheric nitrogen into a form that plants can utilize. The nitrogen fixation process is crucial for preserving soil fertility.

4. Disease Suppression: By competing with pathogenic organisms for resources or by generating substances that stop their growth, certain soil microbes can suppress plant illnesses. There may be less need for chemical pesticides as a result of this natural disease suppression.

5. Improving Soil Structure: By generating compounds that bind soil particles together to form aggregates, soil microbes can also aid in the improvement of soil structure. The infiltration of water, root penetration, and soil aeration are all enhanced by these aggregates.

6. Improving Nutrient Availability: By releasing enzymes that break down organic matter and mineralize nutrients, soil microbes can improve the availability of nutrients to plants. Nutrient uptake by plants is facilitated by this process.

Elements Affecting Microorganisms in the Soil

The quantity and activity of soil microorganisms can be influenced by some factors, including:

1. Soil Moisture: The majority of soil microorganisms prefer moist environments, and soil moisture has an impact on their activity. On the other hand, too much moisture can cause waterlogging, which lowers microbiological activity.

2. Soil pH: Because various microorganisms have varying pH preferences, soil pH affects the sorts of microbes that are present in the soil. The majority of bacteria favor neutral or slightly acidic soils (pH 6.0–7.0).

3. Soil Temperature: The majority of soil microorganisms are more active in warmer temperatures, and soil temperature has an impact on their activity. Extreme heat, however, has the potential to suppress microbial activity.

4. Content of Organic Matter: Soil microbes utilize organic matter as a source of nutrition and energy. Both microbial biomass and activity are often higher in soils that contain a significant amount of organic matter.

5. Oxygen Availability: To breathe, soil microbes need oxygen. Higher levels of microbial activity are supported by well-aerated soils, whereas lower levels of aeration may inhibit microbial activity.

6. Nutrient Availability: For growth and metabolism, soil microorganisms need nutrients like carbon, nitrogen, and phosphorus. Soils that are richer in nutrients typically have more microbial biomass and activity.

For the soil to be healthy and the ecosystem to be sustainable, soil microorganisms are essential. They contribute to the breakdown of organic matter, the cycling of nutrients, the prevention of illness, and the enhancement of soil structure. Comprehending the functions of soil microorganisms and the variables that impact their quantity and functionality is crucial for environmentally friendly soil management techniques.

Farmers may increase soil fertility, support plant development, and strengthen ecosystem resilience by cultivating a healthy soil microbiome.

Erosion of Soil and Preservation

The health of the soil and the sustainability of agriculture are seriously threatened by soil erosion. It happens when water, wind, or other forces carry soil away from its original site. In addition to reducing soil fertility, crop yields, and environmental damage, soil erosion can cause the loss of topsoil, which is rich in nutrients and organic matter. To stop soil erosion and preserve the health of the soil, soil conservation measures are crucial. The origins, effects, and conservation techniques for soil erosion will all be covered here.

Why Soil Erosion Occurs

1. Water Erosion: Raindrop impact and water flowing over the soil's surface are the main causes of water erosion. It can take many different forms, such as gully, rill, and sheet erosion. Steep slopes, intense rainfall, and inadequate soil management techniques are some of the variables that frequently aggravate water erosion.

2. Wind Erosion: When strong winds blow over the soil's surface, they pick up and transport away soil particles. This is known as wind erosion. Arid and semi-arid areas with little or no vegetation cover are more likely to experience wind erosion. It may cause the fertility of the soil to deteriorate and the topsoil to be lost.

3. Tillage Erosion: Tillage erosion is the result of mechanical tillage techniques like harrowing and plowing disturbing the soil. These methods have the potential to dissolve soil aggregates, expose soil to agents that cause erosion, and quicken soil erosion. Tillage erosion can be lessened by using conservation tillage techniques.

4. Deforestation: By eliminating the vegetation cover that shields the soil from agents that cause erosion, deforestation can cause soil erosion. Soil erosion by wind and water is more likely to occur in the absence of vegetation. Erosion is made worse by deforestation because it lowers the soil's organic matter level.

5. Overgrazing: By reducing vegetative cover and compacting the soil, excessive grazing by cattle can cause soil erosion. Because compacted soils allow less water to pass through, runoff and erosion are enhanced. Reducing soil erosion and preventing overgrazing are two benefits of good grazing management techniques.

Effects of Soil Degradation

1. Loss of Soil Fertility: The topsoil, which is rich in nutrients and organic matter and necessary for plant growth, is removed via soil erosion. Eroded soils are hence less fertile and yield fewer crops.

2. Lower Crop Yields: Because of topsoil loss, lowered soil fertility, and increased soil compaction, soil erosion can result in lower crop yields. Additionally, vulnerable to drought and flooding, eroded soils harm crop productivity.

3. Environmental Degradation: The loss of biodiversity, deterioration of water quality, and disturbance of ecosystem processes are all consequences of soil erosion. Additionally, degraded soils can produce sediment runoff that clogs streams and damages aquatic ecosystems.

4. Economic Losses: There might be a lot of financial implications from soil erosion, such as decreased agricultural output, higher soil conservation measures expenditures, and infrastructure damage from sediment runoff.

Techniques for Preserving Soil

1. Contour Plowing: Rather than plowing up and down hills, contour plowing is plowing along the land's contour lines. This slows down the flow of water across the ground, which helps to lessen soil erosion and water runoff.

2. Terracing: Using earthen embankments, terracing entails leveling down steep slopes. Terraces provide a sequence of steps to slow down water runoff and shorten the slope, both of which serve to decrease soil erosion.

3. Cover Cropping: When the primary crop isn't growing, cover crops like legumes or grasses are sown in its place. In addition to improving soil structure and preventing soil erosion, cover crops enrich the soil with organic matter.

4. Conservation Tillage: To reduce soil disturbance, conservation tillage entails lowering the intensity of tillage operations. This promotes improved soil health, decreases soil erosion, and maintains soil structure.

5. Mulching: To prevent soil erosion, a layer of organic or inorganic material, such as plastic, wood chips, or straw, is spread over the soil. Mulches enhance soil moisture retention, lessen runoff, and shield the soil from compaction.

6. Agroforestry: This technique entails incorporating bushes and trees into agricultural environments. Enhancing soil fertility, lowering soil erosion, and giving farmers alternative revenue streams are all possible with agroforestry.

The health of the soil and the sustainability of agriculture are seriously threatened by soil erosion. It may result in diminished crop yields, diminished soil fertility, topsoil loss, and environmental deterioration. To stop soil erosion and preserve soil health, soil conservation techniques including mulching, agroforestry, cover crops, conservation tillage, and contour plowing are crucial. Farmers may enhance crop productivity, safeguard their soils, and advance sustainable agriculture by putting these methods into practice.

Chapter 5: Examining Soils

The Value of Soil Examination

A vital part of sustainable agriculture is soil testing, which yields significant data regarding the pH level, nutrient status, and other aspects of the soil. Through better agricultural yields and environmental sustainability, farmers may make more educated decisions about soil management techniques, crop selection, and fertilization with the use of soil testing. The significance of soil testing, its advantages, and the procedures for carrying it out will all be covered in this chapter.

The Value of Soil Examination

1. Optimizing Fertilizer Use: Farmers can identify any deficits in their soil's nutrient content by conducting soil tests. Farmers may apply fertilizers more effectively, cutting costs and avoiding environmental effects, by understanding the nutrient state of the soil.

2. Increasing Crop Yields: Farmers can adjust fertilizer applications to match the unique nutrient requirements of their crops by conducting soil tests. Profitability may rise along with enhanced agricultural yields and quality as a result of this.

3. Safeguarding the Environment: Overuse of fertilizers can cause nutrient runoff, which contaminates rivers and damages aquatic habitats. Farmers can lower the danger of nutrient pollution and safeguard the environment by optimizing fertilizer use through soil testing.

4. Sustaining Soil Health: By giving farmers information on pH levels, organic matter content, and other crucial parameters, soil testing assists farmers in keeping an eye on the condition of their soil. This enables farmers to gradually increase the fertility and health of their soil.

5. Directing Soil Management Practices: Crop rotation, cover crops, and tillage are a few examples of soil management techniques that can be directed by the useful information that soil testing offers. Farmers can manage their land more effectively if they are aware of the nutrient state of their soil.

6. Economic Benefits: By assisting farmers in avoiding needless fertilizer applications, soil testing might result in economic savings for them. Farmers may enhance their bottom line and save input costs by applying fertilizers more effectively.

The Advantages of Soil Analysis

1. Nutrient Management: By ensuring that crops have access to the nutrients they require for optimum growth, soil testing assists farmers in managing nutrients more effectively.

2. pH Adjustment: By using soil testing, farmers can find out if their soil is excessively acidic or alkaline and receive guidance on how to bring the pH down to a level that is more conducive to plant growth.

3. Increased Crop Yields: Soil testing can increase crop yields and quality by optimizing pH and nutrient levels.

4. Environmental Protection: Farmers may lessen the impact of agriculture on the environment by using soil testing to prevent nutrient runoff.

5. Cost Savings: By preventing needless applications of fertilizer and other inputs, soil testing can help farmers save money.

6. Long-Term Soil Health: Soil testing can assist farmers in preserving and enhancing the long-term health of their soil by tracking its condition over time.

How to Perform a Soil Test

1. GATHERING SOIL SAMPLES: To guarantee a representative sample, soil samples should be gathered from multiple sites within the field. Before testing, all samples have to be taken at the same depth and well mixed.

2. GETTING READY FOR TESTING: Before testing, soil samples need to be air-dried and sieved to get rid of any debris. After that, the samples ought to be packaged and delivered to a lab that tests soil for examination.

3. EXAMINING SOIL SAMPLES: Vital information such as pH, organic matter content, and nutrient content are examined in soil samples. Typically, the findings are given in a report on soil testing that also offers management suggestions.

4. RESULTS INTERPRETATION: After carefully reading the soil test report, farmers should consider how the results relate to their particular farming methods and crop requirements. The report often contains recommendations for soil management techniques including fertilizer application.

5. PUTTING RECOMMENDATIONS INTO EFFECT: To enhance soil health and crop yields, farmers can put recommendations for fertilizer application, pH correction, and other soil management techniques into effect based on the findings of the soil test.

Farmers can benefit greatly from soil testing since it offers crucial information on the fertility and health of the soil. Farmers may increase crop yields, preserve long-term soil health, safeguard the environment, and optimize fertilizer use with the aid of soil testing. Farmers may improve the sustainability and profitability of their enterprises by regularly testing their soil and putting the advice into practice.

How to Gather and Analyze Soil Sample Data

An essential technique in agriculture, soil testing offers important information on the pH balance, nutrient levels, and overall health of the soil. Making educated judgments regarding crop selection, soil management techniques, and fertilization requires accurate soil sample and result interpretation. This chapter will examine the methods for gathering and analyzing soil samples, providing detailed instructions, and addressing important factors.

1. GATHERING SAMPLES OF SOIL

Accurate soil test findings depend on the collection of representative soil samples. This is a detailed how-to for soil sampling:

Step 1: Choose the Areas for Sampling

- Locate the sections of your garden or field that best illustrate various soil types, past agricultural methods, or crop rotations.
- Using criteria including topography, soil type, and past land use, divide the area into uniform sample zones.

Step 2: Choose Sampling Tools.

- To gather soil samples, use a clean sampling instrument such a spade, auger, or soil probe.

- To prevent contamination, make sure the sampling tool is composed of stainless steel or another non-reactive material.

Step 3: **Choose the Sampling Depth**

- Based on the root zone of the crops you plan to produce, determine the proper sampling depth.
- A depth of 6 to 8 inches should be sampled for the majority of agricultural crops. To sample trees and shrubs, go 12 to 18 inches down.

Step 4: **Gathering Samples**

- Within each survey zone, gather soil samples at random; stay away from regions that clearly differ, such as compost piles or fertilizer.
- To ensure complete coverage, gather samples across the sampling area in a zigzag or W-shaped manner.
- To create a representative composite sample, gather many subsamples from each sampling zone at regular intervals.

- To form a composite sample for analysis, place each subsample in a clean bucket or container and thoroughly mix them.

Step 5: **Sample Labeling and Storage**

- Affix a unique identification to every sample, together with details about the location, time, and sampling depth.
- To avoid contamination and maintain soil moisture, store samples in sterile, airtight containers.
- As samples are being transported to the testing facility, keep them cold and out of direct sunlight.

2. INTERPRETING THE RESULTS OF SOIL TESTING

Understanding the fundamentals of soil fertility and chemistry is necessary to interpret the findings of soil tests. This is how typical soil test parameters should be interpreted:

1. pH Scale

- ✓ Soil pH determines the soil's acidity or alkalinity and affects the availability of nutrients to plants.
- ✓ Crop-specific optimal pH ranges vary, but most plants prefer slightly acidic (pH 6.0–7.0) soils over neutral ones.
- ✓ Use elemental sulfur to lower pH in alkaline soils or lime to boost pH in acidic soils as needed.

2. NPK macronutrients

- ✓ Soil testing determine the concentrations of nitrogen (N), phosphorus (P), and potassium (K), three important macronutrients.
- ✓ Interpret nutrient levels according to recommendations for soil fertility and crop requirements.
- ✓ Modify fertilizer treatments as necessary to preserve soil fertility and fulfill crop nutrient requirements.

3. Tiny Nutrients

- ✓ Micronutrients including iron (Fe), manganese (Mn), zinc (Zn), and copper (Cu) can also be examined in soil tests.
- ✓ Crop-specific requirements for micronutrients differ, and plants may suffer from excess micronutrient levels.
- ✓ Apply micronutrient amendments as directed by the results of a soil test.

4. Content of Organic Matter

- ✓ The organic matter in the soil has a role in the nitrogen cycle, soil structure, and moisture retention.
- ✓ A higher percentage of organic matter is generally better for the fertility and health of the soil.
- ✓ To increase the amount of organic matter in the soil, add organic amendments like manure or compost.

5. Capacity Exchange Cation (CEC)

- ✓ CEC gauges how well soil holds and exchanges cations like potassium, magnesium, and calcium.
- ✓ Greater nutrient retention ability and buffering against nutrient leaching are indicated by higher CEC values.
- ✓ To maximize nutrient availability and reduce losses, modify fertilizer treatments in accordance with CEC values.

6. Guidelines for Interpretation

- ✓ Consult the guidelines for interpreting soil test results that the testing laboratory or agricultural extension service have supplied.
- ✓ Seek advice from agronomists or soil scientists for ideas that are particular to your crop and soil type.
- ✓ To monitor changes in soil fertility over time, maintain thorough records of management decisions and soil test findings.

Advice on How to Sample and Interpret Soil

- ❖ **Sampling Frequency:** To track changes in soil health and fertility over time, soil sampling should be done on a regular basis—at least once every two to three years.
- ❖ **Depth of Sampling:** Take soil samples at the right depth for your crops. Samples from the top 6 to 8 inches of soil should be obtained for the majority of crops.
- ❖ **Composite Sampling:** Combine several subsamples to create composite samples. A better representative sample of your soil will result from doing this.
- ❖ To maximize soil health and crop production, **heed the advice given in your soil test result regarding fertilization,** lime application, and other soil management techniques.

A vital component of soil testing and nutrient management in agriculture is gathering and analyzing soil samples.

Farmers can decide on fertilization, soil amendments, and crop management techniques by following the right sampling protocols and comprehending the findings of soil tests. Farmers may maintain soil health, maximize fertilizer use efficiency, and accomplish sustainable agricultural production with the help of soil testing. A proactive soil management plan that maximizes crop yields while minimizing environmental impacts must include routine soil testing and interpretation.

Utilizing Soil Test Findings for Management of Soil

Farmers and gardeners can evaluate the fertility and health of their soils with the help of soil testing. They can ascertain pH, nitrogen levels, and other elements that influence plant growth by examining soil samples. Making educated decisions about fertilization, soil amendments, and general soil management techniques is made possible by the interpretation of soil test data. We will look at how to successfully use soil test findings for soil management here.

1. COMPREHENDING SOIL TEST REPORTS

Reports from soil tests give specific details about the pH and nutrient content of the soil. It is vital to comprehend these reports in order to make well-informed decisions on soil management. The following are the main elements of a standard soil test report:

- ❖ **Nutrient Levels:** Soil test reports will list the amounts of macronutrients (potassium, phosphorus, and nitrogen) as well as micronutrients (zinc, iron, manganese, and so forth). Usually, these amounts are expressed in pounds per acre or parts per million (ppm).
- ❖ **pH Level:** The soil's pH level reveals how acidic or alkaline the soil is. The majority of plants like a pH of 6.0 to 7.0, which is slightly acidic to neutral. The

pH level of the soil sample will be indicated in soil test reports.

- ❖ **Organic Matter Content:** The fertility and health of the soil depend on the organic matter in the soil. The proportion of organic matter content can be seen in soil test data.
- ❖ **Cation Exchange Capacity (CEC):** The soil's capacity to retain and exchange nutrients is gauged by its Cation Exchange Capacity (CEC). Higher CEC soils are better at retaining nutrients.
- ❖ **Interpretation and Recommendations:** Reports on soil tests frequently provide an interpretation of the findings along with suggestions for soil management techniques like pH correction and fertilization.

2. UTILIZING FERTILIZATION RESULTS FROM SOIL TESTS

Determining the soil's need for fertilizer is one of the main applications of soil test results. Here's how to fertilize using the findings of a soil test:

- ❖ **Nutrient Deficiencies:** The results of a soil test will show whether the soil is lacking in any particular nutrients. Farmers can use fertilizers containing the insufficient nutrient if one has been discovered.
- ❖ **Nutrient surpluses:** On the other hand, data from soil tests can point to nutrient surpluses. Plants may suffer from excessive nutrient levels, in which case remedial measures like lowering fertilizer applications may be necessary.
- ❖ **Fertilizer Selection:** Farmers can choose fertilizers that satisfy the unique nutrient requirements of their crops based on the findings of soil tests. This lowers expenses and helps avoid applying fertilizer excessively.

- **Application Rates:** The findings of soil tests can also be used to establish the proper fertilizer application rates. Farmers can maximize nutrient use efficiency and reduce environmental effects by applying fertilizers at the proper rates.

3. CHANGING THE PH OF THE SOIL

The pH of the soil is a major factor in the nutrients that plants can get. The pH level of the soil will be indicated by the results of a soil test, and changes may be required to bring the pH level to the ideal range for plant growth. Here's how to change the pH of soil according to the findings of a soil test:

- **Acidic Soils:** Farmers can apply lime to raise pH levels if soil test results show that the soil is excessively acidic (below the recommended pH range). Lime reacts with the soil to neutralize acidity.

- ❖ **Alkaline Soils:** Farmers can use elemental sulfur to lower pH levels if soil test results show that the soil is excessively alkaline, or above the ideal pH range. Elemental sulfur reacts with the soil to produce an acidic pH.
- ❖ **Maintaining pH:** Farmers can keep an eye on soil pH levels and make necessary adjustments to maintain the ideal pH for plant growth by regularly testing their soil.

4. KEEPING AN EYE ON THE HEALTH OF THE SOIL

The results of soil tests can be used to track changes in the soil over time and offer important information about its health. Farmers can monitor soil fertility, organic matter content, and other critical soil factors by regularly testing their soil. Proactive soil management techniques to preserve soil health and productivity are made possible by this information.

5. ENHANCING THE STRUCTURE AND DRAINAGE OF SOIL

The findings of soil tests might also shed light on drainage problems and soil structure. Plant growth can be adversely affected by issues with soil structure, inadequate drainage, and compaction. Farmers can enhance soil structure and drainage, resulting in healthier soils and increased crop yields, by correcting these problems in accordance with the findings of soil tests.

Methods of Sustainable Soil Management

Sustainable agriculture places a strong emphasis on managing the soil by using the results of soil testing. Through the optimization of nutrient utilization, pH adjustment, and soil health enhancement, farmers can mitigate environmental effects, preserve resources, and foster soil productivity over an extended period. A crucial element for sustainable soil management techniques is soil testing.

Create a Plan for Soil Management

- ❖ Create a thorough soil management plan that takes into account fertilizer requirements, pH adjustments, and organic matter management based on the findings of soil tests.
- ❖ Take into account in your plan elements like cover crops, crop rotation, and tillage techniques.

The findings of soil tests are useful in directing soil management procedures. Farmers and gardeners can make well-informed decisions about fertilization, pH correction, and general soil health by comprehending and interpreting soil test findings. A proactive method to managing soil is soil testing, which makes sustainable measures that support crop productivity and soil health possible. It's advised to conduct routine soil tests to keep an eye on soil fertility and make necessary adjustments to ensure the best possible soil conditions for plant growth.

Chapter 6: Methods of Soil Management

Soil Amendments: Organic vs. Inorganic

Enhancing the fertility, structure, and general health of the soil requires the application of soil amendments. Different benefits and factors for managing soil apply to inorganic and organic soil additions. This chapter will examine the distinctions between inorganic and organic soil amendments, as well as the benefits and drawbacks of each, and how to select the best amendments for the requirements of your crop and soil.

1. AMENDMENTS TO ORGANIC SOIL

Compost, manure, and plant leftovers are examples of natural sources from which organic soil additions are made. These ingredients include nutrients and organic particles that support microbial activity and soil fertility. The following are a few typical organic soil amendments:

- ❖ **Compost:** Compost is organic matter that has broken down and is full of microbes and nutrients. It gives plants a slow-release supply of nutrients and strengthens the structure of the soil while retaining more water.
- ❖ **Manure:** An excellent supply of organic matter, potassium, phosphate, and nitrogen is animal dung. It boosts nitrogen cycling, microbial activity, and soil fertility and structure.
- ❖ **Cover Crops:** Cover crops are cultivated specially to increase the fertility and health of the soil. They keep weeds at bay, replenish the soil with organic materials, and stop erosion. Legumes (such as clover, peas) and grasses (such as rye, oats) are common types of cover crops.

Benefits of Adding Organic Matter to Soil

Better Soil Structure: By enhancing aggregate stability and lowering compaction, organic soil additions enhance soil structure.

Enhanced Water Retention: The addition of organic matter to soil reduces water runoff and increases drought tolerance by improving water retention.

Nutrient Recycling: By releasing nutrients gradually as they break down, organic soil amendments recycle nutrients and lessen the demand for synthetic fertilizers.

Promotion of Beneficial Microorganisms: The microbial community in the soil is made more diversified by organic amendments, which improves disease prevention, nutrient cycling, and general soil health.

The drawbacks of adding organic matter to soil

Slow Release of Nutrients: As organic soil additions break down, nutrients are released gradually, which may not be enough to meet crops' immediate nutrient needs.

Potential for Contamination: If organic soil additives like manure are not adequately composted or treated, they may harbor diseases or weed seeds.

Variable Nutrient Content: A number of variables, including the source material, the composting process, and the storage conditions, can affect the nutrient content of organic soil additions.

2. INORGANIC MODIFICATIONS TO SOIL

Synthetic or mineral fertilizers, commonly referred to as inorganic soil amendments, are made of inanimate substances like chemicals, salts, and minerals. They can be tailored to satisfy individual nutrient requirements and give plants easily accessible nutrients. Typical forms of inorganic soil additions include the following:

- ❖ **Nitrogen Fertilizers:** Nitrogen is made available to plants in a form that they can easily absorb by the use of nitrogen fertilizers like urea, ammonium nitrate, and ammonium sulfate. They encourage plants to grow vegetatively and to turn green.
- ❖ **Phosphorus Fertilizers:** Phosphorus is supplied in a form that plants may easily absorb by phosphorus fertilizers like superphosphate and triple superphosphate. They encourage the growth of roots, flowers, and fruits.
- ❖ **Potassium Fertilizers:** Potassium is supplied in a soluble form for plant uptake via potassium fertilizers, such as potassium sulfate and chloride. They support fruit quality, disease resistance, and general plant health.

Benefits of Inorganic Amendments to Soil

Instant Nutrient Availability: Inorganic fertilizers are good for addressing nutrient deficits and accelerating plant growth since they provide plants access to nutrients right away.

Accurate Nutrient Content: Because inorganic fertilizers are known to contain specific nutrients, they may be applied at exact rates and with focused nutrient management.

Convenience: Large-scale agricultural enterprises find inorganic fertilizers convenient due to their ease of application, storage, and transportation.

The drawbacks of adding inorganic materials to soil:

Potential for Nutrient Leaching: If inorganic fertilizers are applied excessively or during periods of high precipitation, they may seep into surface or groundwater, causing nutrient contamination and environmental damage.

Soil Acidification: Over time, some inorganic fertilizers, including those based on ammonium, may cause the soil to become more acidic, which can result in pH imbalances and nutritional deficits.

Harmful Effect on Soil Microorganisms: If inorganic fertilizers are misused or administered improperly, they can damage soil microbial communities and decrease soil biodiversity.

3. SELECTING APPROPRIATE SOIL AMENDMENTS

The selection between organic and inorganic soil amendments is contingent upon a number of criteria, such as crop requirements, soil type, nutrient availability, and environmental factors. The following criteria can help you select the appropriate soil amendments:

- **Soil Test:** To evaluate the pH, nitrogen levels, and other characteristics of the soil, conduct a soil test. Finding out which nutrients are lacking and if organic or inorganic amendments are required can be aided by the findings of soil tests.
- **Crop Requirements:** Take into account the nutrients that your intended crops will need. The choice of soil amendments may be influenced by the higher concentrations of certain nutrients that some crops may demand.
- **Environmental Impact:** Take into account how soil amendments may affect soil erosion, nutrient runoff, and leaching. Select soil additions that support sustainable soil management techniques and reduce environmental concerns.
- **Cost and Availability:** Take into account the affordability and accessibility of soil amendments in addition to their efficacy and simplicity of use. Select the amendments that offer the best value for the needs of your particular crop and soil.

4. EMPLOYING A MIXTURE OF AMENDMENTS

The best method of managing soil in many circumstances might involve combining inorganic and organic amendments. For instance, adding organic materials to the soil, like compost or manure, can increase its fertility and structure. Meanwhile, adding inorganic fertilizers as a supplement might help the soil retain certain minerals that may be deficient. The advantages of both kinds of changes can be enhanced with the aid of this integrated strategy.

Soil fertility and management are significantly impacted by both organic and inorganic soil additions. While inorganic supplements supply easily accessible nutrients for quick plant growth, organic amendments strengthen soil structure, improve water retention, and encourage microbial activity. A number of variables, including soil type, crop requirements, nutrient availability, and environmental concerns, influence the decision between organic and inorganic supplements. Farmers and gardeners may enhance soil health and maximize crop productivity by making educated selections by knowing the benefits and drawbacks of each type of amendment.

Using Compost to Improve Soil

Organic materials are naturally converted through the process of composting into compost, a nutrient-rich soil additive. Enhancing soil fertility, structure, and general health can be accomplished efficiently and sustainably by composting. The advantages of composting, the technique of composting, and the application of compost for soil enhancement will all be covered here.

Advantages of Composting

For the fertility and health of the soil as well as the sustainability of the ecosystem, composting has several advantages. Among the main advantages of composting are:

Better Soil Structure: Compost strengthens the soil's aggregate stability, which facilitates root penetration and water infiltration.

Nutrient Enrichment: In addition to micronutrients, compost is high in organic matter, nitrogen, phosphorus, and potassium. These nutrients are gradually delivered over time, giving plants a consistent source to thrive.

Enhanced Microbial Activity: Beneficial microorganisms found in compost are abundant and aid in the breakdown of organic materials, the release of nutrients, and the suppression of plant diseases.

Decreased Soil Erosion: By enhancing soil stability and structure, compost reduces soil erosion.

Environmental Benefits: By keeping organic waste out of landfills, composting lowers greenhouse gas emissions and lessens the demand for chemical fertilizers.

The Process of Composting

The natural process of composting entails the aerobic decomposition of organic materials by microbes. There are various phases to the composting process:

Stage 1: Initial Decomposition: Sugars and starches, which are easily broken down by microbes, are broken down in this stage. One consequence of microbial activity is heat.

Stage 2: Active Decomposition: More complex organic components like cellulose and lignin are broken down when the compost pile heats up. Rapid breakdown and intense microbial activity are characteristics of this stage.

Stage 3: Curing: During this phase, the decomposition process slows down and the compost pile cools. Organic matter is still broken down by microorganisms, although at a slower pace.

Stage 4: Maturity: When compost has an earthy scent and a crumbly, dark texture, it is said to be mature. Compost that has reached maturity can be added to soil since it is stable.

Methods for Composting

Composting can be done in a variety of ways, each with pros and cons. Typical techniques for composting include:

Bin Composting: This method entails keeping compost materials contained within a bin or enclosure. This technique works well for small-scale composting and works with a range of materials, including metal, plastic, and wood.

Pile Composting: This method entails building a sizable heap of compost on the ground. Larger amounts of compost materials can be used using this method, which also permits natural aeration and decomposition.

Vermicomposting: In this method, organic materials are broken down by worms. This approach yields high-quality compost and is appropriate for small-scale composting and kitchen waste.

Windrow Composting: Windrow composting is building compost materials into long, narrow stacks. This technique is frequently applied in agricultural contexts and is appropriate for large-scale composting.

Using Compost to Improve Soil

There are several methods to use compost to enhance the fertility, structure, and general health of the soil. Typical applications for compost include:

Soil Amendment: You can add compost to your soil to enhance its nutrient content, water retention, and structure. Composted soil has higher microbial activity, which improves soil health and nutrient cycling.

Mulching: To cover the soil's surface, compost can be utilized as a mulch. Compost mulching aids in controlling soil temperature, weed suppression, and moisture retention.

Compost Tea: Compost is steeped in water to create compost tea, a liquid fertilizer. Because it is full of microorganisms and nutrients, compost tea is a great foliar spray and soil drench for plants.

Topdressing: You can give established plants a layer of compost on top of them. Compost topdressing enriches the soil with nutrients and strengthens its structure without uprooting plant roots.

The Best Practices for Composting

Observe these best procedures to guarantee a good composting process:

Maintain a balanced proportion of carbon-rich materials (such as leaves and straw) to nitrogen-rich materials (such as kitchen trash and grass clippings) for the best possible breakdown.

Sufficient Aeration: To ensure that the compost pile receives enough oxygen and air, turn it frequently. This facilitates the breakdown process's acceleration.

Control Moisture: Make sure the compost pile is damp but not soggy. Like a damp sponge, the perfect moisture level is reached when.

Temperature Monitoring: Keep a frequent eye on the compost pile's temperature. Heat is produced by an actively decaying compost pile, which is essential for the breakdown of organic matter.

A useful technique for managing soil that enhances soil fertility, structure, and general health is composting. Compost is a great soil supplement for landscapes, farms, and gardens since it's full of nutrients, organic matter, and helpful microbes. Farmers and gardeners can make use of composting's advantages to enhance soil quality and advance sustainable agriculture by being aware of the process, using compost wisely, and adhering to best practices.

Crop rotation and cover cropping

Two fundamental soil management techniques that are critical to sustainable agriculture are cover crops and crop rotation. These methods lessen erosion and nutrient leaching while enhancing the fertility, health, and structure of the soil. Here we will discuss the advantages of crop rotation and cover crops, as well as their application and effects on soil health and crop yield.

Cover Cropping

Planting a transient, non-harvested crop to cover the soil's surface is known as cover cropping. Numerous advantages for the fertility and health of the soil come from cover crops. Legumes (such as vetch, and clover), grasses (such as rye, and oats), and brassicas (such as mustard, and radish) are a few examples of common cover crops.

The advantages of cover crops

Soil Erosion Control: By lessening the effect of rains and enhancing soil structure, cover crops help to prevent soil erosion.

Weed Suppression: By competing with weeds for nutrients, sunshine, and moisture, cover crops lessen the amount of weed pressure in the field.

Nutrient cycling: When cover crops are absorbed into the soil or left as mulch, the nutrients they absorb from the soil are released back into it.

Better Soil Structure: By breaking up compacted soil and forming pathways for air and water movement, cover crop roots contribute to a better soil structure.

Increased Soil Organic Matter: The organic matter that covers crops add to the soil during their decomposition increases microbial activity and soil fertility.

Putting Cover Cropping into Practice:

Choosing Cover Crops: Take into account your objectives for improving the soil as well as the unique soil and climate conditions in your area when selecting cover crops. Take into account elements like biomass production, weed control, and nitrogen fixation.

Planting Cover Crops: Considering your climate and location, plant cover crops at the right time. For germination, make sure the seed has good soil-to-seed contact and enough moisture.

Managing Cover Crops: Keep an eye on the development of your cover crops and adjust as necessary. While certain cover crops can be allowed to grow until they naturally senesce, others might need to be stopped before cash crops are planted.

Including Cover Crops: Use tillage or mowing to include cover crops in the soil. Permit the residues from the cover crops to break down and improve the soil.

Crop Rotation

Crop rotation is the practice of growing various crops on the same plot of land in a predetermined order. Crop productivity and soil health both benefit from crop rotation in a number of ways.

The advantages of crop rotation

Disease and Pest Management: By disrupting the life cycles of pests and diseases that target certain crops, crop rotation can aid in breaking the cycles of disease and pests.

Nutrient Management: The needs for certain nutrients vary throughout crops. Crop rotation lowers the possibility of nutrient imbalances and enables the soil to use nutrients more effectively.

Weed Suppression: By upsetting weed life cycles and lowering weed seed banks in the soil, crop rotation can aid in the suppression of weeds.

Better Soil Structure: The root systems of certain crops can contribute to better soil structure. Crops with fibrous roots can aid in enhancing soil tilt, while deep-rooted crops can loosen up compacted soil.

Crop Rotation Implementation

Organizing Crop Sequences: Arrange crop sequences according to the particular requirements of your land and crops. Take into account elements including nutrient needs, disease and pest challenges, and weed control techniques.

Rotating Crop Families: To lower the chance of a buildup of pests and diseases, rotate crops from several plant families. Do not repeatedly plant the same crop in the same field, or ones that are closely similar.

Cover Cropping as a Component of Crop Rotation: To further enhance soil fertility and health, use cover crops into your crop rotation strategy. Make use of cover crops to help achieve your aims for soil management and cash crops.

Monitoring and Modifying: Keep an eye on how crop rotation affects crop productivity and soil health. Based on your observations and the findings of your soil test, modify your crop rotation plan as necessary.

Two crucial soil management techniques that can enhance soil fertility, health, and overall agricultural productivity are cover crops and crop rotation. Farmers can improve soil structure, control pests and diseases, manage weeds, and lessen soil erosion by implementing these strategies into their farming operations. Two essential elements of sustainable agriculture that support soil health and environmental care are cover crops and crop rotation.

Chapter 7: Classification and Types of Soils

An Overview of Systems for Classifying Soil

The properties and characteristics of soil vary greatly, making it a dynamic and complicated natural resource. A methodical approach to classifying soils according to their physical, chemical, and mineralogical characteristics is called soil classification. Engineering, environmental management, land use planning, and agriculture all depend on an understanding of soil kinds and classification. The many kinds of soil and the several classification schemes that are employed to group them will be discussed in this chapter.

Types of Soil

Based on particle size, there are three main types of soil: silt, clay, and sand. The components of soil include these particles, organic matter, air, and water. The texture and characteristics of the soil are determined by the relative amounts of these particles.

SAND: With sizes ranging from 0.05 to 2.0 millimeters, sand particles are the largest of the three categories. Although sandy soils drain easily, they struggle to hold onto moisture and minerals.

SILT: With sizes ranging from 0.002 to 0.05 millimeters, silt particles are smaller than sand particles. Although silty soils retain water well, they are quickly compacted.

CLAY: With dimensions less than 0.002 millimeters, clay particles are the smallest of the three categories. Although clay soils retain a lot of water and nutrients, they can also be prone to compaction and have poor drainage.

Systems for Classifying Soils

A methodical approach to classifying soils according to their physical, chemical, and mineralogical characteristics is called soil classification. Land use planning, engineering, environmental science, agriculture, and other fields depend on standardized frameworks for understanding and expressing soil characteristics, which are provided by soil categorization systems. An overview of soil categorization systems, such as the World Reference Base for Soil Resources (WRB), the USDA soil taxonomy, and various regional systems, will be given in this chapter.

1. USDA TAXONOMY OF SOILS

In the US and many other nations, the USDA soil taxonomy is the most extensively utilized system for classifying soils. Based on their physical, chemical, and mineralogical characteristics, it divides soils into tiers.

There are six hierarchical levels in the USDA soil taxonomy, which include:

- ❖ **ORDER:** The most advanced classification scheme, determined by factors like soil form, temperature, and moisture content.
- ❖ **SUBORDER:** A separation of soil orders according to extra characteristics of the soil, like mineralogy, texture, and structure.
- ❖ **GREAT GROUP:** A subset of soil suborders distinguished by more particular characteristics like color, texture, and structure.
- ❖ **SUBGROUP:** An additional segmentation of soil great groups according to more specific soil characteristics.
- ❖ **FAMILY:** A division of soil subgroups according to particular characteristics of the soil, like the features of the soil horizon.
- ❖ **SERIES:** The most granular classification level, determined by distinct soil characteristics including parent material and placement in the landscape.

More specific information about the soil is provided at each classification level, making it easier to identify various soil types and comprehend their characteristics.

2. THE SOIL RESOURCES WORLD REFERENCE BASE (WRB)

The International Union of Soil Sciences (IUSS) created the global soil classification system known as the World Reference Base for Soil Resources (WRB). The WRB is dependent on various soil variables, including parent material, soil formation processes, and characteristics of the soil profile. Based on the primary soil processes and features, the WRB divides soils into 32 reference soil categories.

In contrast to the USDA soil taxonomy, the WRB is more adaptable and can take into account a larger variety of soil types and characteristics. The WRB is used as a standard for soil categorization in several nations and is intended to be universally applicable.

3. ADDITIONAL REGIONAL SYSTEMS FOR CLASSIFYING SOILS

Many nations and regions have their own soil classification systems that are adapted to their unique soil types and qualities, in addition to the USDA soil taxonomy and the WRB. These regional methods are better suited for local soil management techniques since they frequently take into account local soil features and categorization standards.

Regional soil classification systems include, for instance:

Canadian System of Soil Classification: This system, which is in use in Canada, divides soils into different categories according to their texture, horizons, and other characteristics.

Australian Soil Classification: This method, which is used in Australia, divides soils into groups according to characteristics including color, texture, and structure.

FAO Soil Classification System: Created by the Food and Agriculture Organization (FAO), this system categorizes soils according to whether or not they are suitable for agriculture. It is utilized in many developing nations.

The Significance of Classifying Soils

Classifying soil is crucial for some reasons.

Understanding Soil Properties: Agriculture, engineering, and environmental research all depend on our ability to comprehend the physical, chemical, and mineralogical characteristics of soils, which is made possible by soil classification.

Land Use Planning: By supplying data on soil fertility, drainage, and other attributes that influence land use suitability, soil classification aids in the making of land use planning decisions.

Environmental Management: By gaining an understanding of how land use patterns like mining, agriculture, and urban growth affect the environment, soil categorization enables us to create sustainable land management strategies.

Soil Conservation: Using soil classification, we may conserve soils that are vulnerable to erosion, compaction, and other types of deterioration by putting conservation measures in place.

Classification Criteria for Soils

Soils are categorized according to many factors, such as:

- ❖ **Physical characteristics:** Soils are categorized based on their texture, structure, and color. These characteristics affect drainage, water retention, and soil fertility.

- ❖ **Chemical Properties:** Soils are also categorized based on their chemical characteristics, which include pH, nutrient content, and cation exchange capacity (CEC). These characteristics affect plant nutrition availability and soil fertility.
- ❖ **Mineralogical Properties:** Soils can also be categorized based on their mineralogical characteristics, such as the presence of iron oxides and clay minerals. These characteristics affect nutrient retention and soil structure.

A methodical approach to classifying soils according to their physical, chemical, and mineralogical characteristics is called soil classification. Standardized frameworks for comprehending and expressing soil characteristics are offered by the WRB, the USDA soil taxonomy, and various regional soil classification systems. For the purpose of understanding soil qualities, guiding land use decisions, and promoting sustainable soil management, soil classification is crucial for the fields of agriculture, engineering, environmental science, and land use planning.

Principal Soil Types in the World

Terrestrial ecosystems are built on soils, which facilitate nutrient cycling, water filtration, and plant growth. Globally, soils exhibit great variation in their characteristics, giving rise to different types of soils depending on variables including terrain, parent material, climate, and time. Planning land use, environmental management, and agriculture all depend on having a thorough understanding of these soil types. This chapter examines the main varieties of soil that are found worldwide along with some of their attributes.

1. MOLLYSOLS

Distribution: Grasslands across the planet, especially in North America, South America, Europe, and Asia, are home to mollusks. They can also be found in some areas of Australia and Africa.

Features: Mollisols are distinguished by their thick, organic-rich surface horizon (topsoil), dark color, and high fertility. They are perfect for agriculture, especially for growing grasses and cereals, as they drain efficiently.

2. ALFISOLS

Distribution: Alfisols can be found worldwide in both tropical and temperate climates. They are widespread over sections of Africa and South America, as well as the mid-latitudes of North America, Europe, and Asia.

Features: The subsurface of alfisols is rich in clay and has a modest level of fertility. Although they drain properly, erosion may occur. Grains, fruits, and vegetables are among the many crops grown in alfisols.

3. SPODOSOLS

Distribution: The main habitats for spodosols are cold, damp areas with coniferous woods, which include sections of Asia, northern Europe, and northern North America.

Features: Spodosols are distinguished by a broad, pale-colored surface horizon (E horizon) that is comprised of iron and aluminum oxides as well as organic materials. They have a nutrient-poor, leached subsoil and are acidic. Although they are not the best for agriculture, spodosols are crucial for forestry.

4. ARIDISOLS

Distribution: Aridisols are found in semi-arid and dry areas of the planet, including as Australia, Africa, the Middle East, South America, and North America.

Characteristics: Aridisols are distinguished by their high calcium carbonate content, low organic matter content, and restricted leaching. They can be salty and are frequently shallow. Although they provide difficulties for agriculture, aridisols can be controlled using irrigation and soil amendments.

5. VERTISOLS

Distribution: Tropical and subtropical areas with distinct wet and dry seasons are home to vertisols. They are widespread in some regions of South America, Africa, India, and Australia.

Features: A high degree of shrink-swell qualities and a high clay content are characteristics of vertisols. Deep fissures in the soil are caused by them; they split when dry and swell when moist. Although they provide challenges for agriculture, vertisols can yield good results when managed well.

6. HISTOSOLS

Distribution: Throughout the planet, marshes and peatlands are home to histosols. They are widespread in areas of North America, Europe, and Asia that receive a lot of rainfall or have inadequate drainage.

Features: Histosols are frequently extremely fertile environments with a thick coating of organic matter (peat). They may be drained and used for agriculture, but doing so may cause subsidence. They are crucial to wetland ecosystems.

7. ANDISOLS

Distribution: Andisols are found in volcanic regions all over the world, including areas of Central and South America, Japan, and the Pacific Northwest of the United States, which are all part of the Pacific Ring of Fire.

Features: Andisols have special qualities like high fertility, high water-holding capacity, and high cation exchange capacity (CEC) due to a high content of volcanic ash and glass. Though they can be prone to erosion, they are perfect for agriculture.

8. GELISOLS

Distribution: Gelisols are distributed in cold climates with permafrost, including the Arctic, Antarctic, and high alpine regions.

Characteristics: Permafrost is present within two meters of the soil surface, which is a characteristic of gelisols. They have a brief growing season and little drainage. Although they are unsuitable for cultivation, gelisols are crucial for comprehending historical climate conditions.

Globally, there is a great variation in the qualities and traits of soil types due to various factors like terrain, parent material, climate, and time. Planning land use, environmental management, and agriculture all depend on having a thorough understanding of these soil types. We can more effectively manage and preserve this essential natural resource for future generations if we are aware of the main types of soil and their properties.

Mapping and Interpretations of Soils

The act of identifying and categorizing various soil types in a certain area is known as soil mapping. In agricultural planning, environmental management, and land use planning, it is an essential stage. Important details regarding soil fertility, characteristics, and appropriateness for different purposes can be found on soil maps. The methods, procedures, and interpretation of soil maps for efficient land management will all be covered here.

Overview of Soil Mapping

To produce maps that show the spatial distribution of various soil types, soil data must be gathered, analyzed, and interpreted. The foundation of soil maps is field-based soil surveys, in which soil samples are gathered and examined for a range of characteristics, including pH, structure, texture, and nutrient content. Following that, these data are utilized to produce intricate maps that illustrate how various soil types are distributed throughout a terrain.

Methods for Mapping Soils

Soil mapping involves the use of several techniques, such as:

Field Surveys: In a field survey, soil samples are gathered and visual evaluations of the qualities of the soil are made. Detailed maps of the soil are made using this information.

Remote Sensing: Large areas may be rapidly and affordably mapped using remote sensing techniques like satellite imaging and aerial photography. These methods can yield important details on the characteristics of the terrain and the patterns of the soil.

Geographic Information Systems (GIS): GIS is a tool for organizing, interpreting, and visualizing data about soil. Detailed maps of soil can be created using GIS software and layered with other spatial data, including topography and land use.

Soil Sampling: This process entails gathering soil samples from various areas within a field or landscape. To ascertain the texture, pH, and nutritional content of the soil, these samples are examined in a laboratory.

Interpretation of Soil Maps

Making informed judgments regarding land use and management requires a grasp of the information provided on soil maps. Among the crucial elements in interpreting soil maps are:

Soil Types: Using colors or patterns, soil maps usually show the many types of soil. Different features and traits, like texture, fertility, and drainage, are linked to different types of soil.

Soil Properties: Data on soil characteristics like pH, organic matter content, and nutrient levels may also be included on soil maps. Farmers and land managers can use this information to assess a soil's appropriateness for a given crop or usage of the land.

Land Use appropriateness: Soil maps can be used to evaluate a soil's appropriateness for a variety of land uses, including forestry, urban development, and agriculture. For instance, fertile, well-drained soils might be good for farming, whereas poorly-drained soils might be more suited for protecting wetlands.

Soil Conservation: Areas vulnerable to soil erosion or deterioration can be found using soil maps. Land managers can conserve and enhance the health of the soil by implementing soil conservation measures based on their understanding of the types and attributes of the soil in these locations.

Uses for Soil Mapping

Among the many significant uses for soil mapping are the following:

AGRICULTURE: Site-specific nutrient management plans are developed using soil maps, which are used to assess a soil's compatibility for a given crop.

ENVIRONMENTAL MANAGEMENT: Sustainable land management plans are developed by utilizing soil maps to evaluate the environmental impact of land use practices.

URBAN PLANNING: Soil maps are used in urban planning to determine development-suitable locations and to evaluate the likelihood of soil-related risks like subsidence and landslides.

Obstacles and Restrictions

The following are some of the difficulties and restrictions that come with soil mapping:

SCALE: Regional or national soil maps are frequently created, which may not adequately represent the diversity of local soils.

DATA AVAILABILITY: Comprehensive soil data is necessary for soil mapping; however, this data may not always be available or current.

INTERPRETATION: The usefulness of soil maps for non-specialists may be limited since they necessitate knowledge of soil science and land management.

A useful technique for comprehending soil characteristics and trends across landscapes is soil mapping. Soil maps include comprehensive data on soil types, characteristics, and appropriateness for various land uses. This information can be used to guide land management decisions and encourage sustainable land use practices. Improvements in GIS, remote sensing, and soil sample methods have increased the accessibility and accuracy of soil mapping, enabling more informed land management choices.

Chapter 8: Sustainable Soil Practices

Techniques for Preserving Soil

Preventing soil erosion and deterioration while preserving soil fertility and health is known as soil conservation. To guarantee agricultural lands' long-term productivity and safeguard the environment, sustainable soil practices are crucial. This chapter will examine several methods for conserving soil, such as managing soil fertility, reducing soil erosion, and enhancing soil health.

An Overview of Soil Preservation

The natural process of soil erosion happens when soil particles get separated and are carried by the wind, water, or other forces. However, soil erosion and deterioration can be accelerated by human activities like agriculture, deforestation, and urbanization. The goal of soil conservation is to maintain the productivity and health of the soil by preventing or minimizing these effects.

Methods for Controlling Soil Erosion

To stop soil erosion and safeguard the health of the soil, several soil erosion management measures can be applied:

CONTOUR PLOWING: In contrast to up-and-down plowing, contour plowing entails plowing across the land's slope. This lessens soil erosion and water discharge.

TERRACING: To slow down water flow and let it seep into the soil, level platforms are built on sloping terrain. By retaining soil moisture, tarmacking aids in preventing soil erosion.

COVER CROPPING: In order to prevent soil erosion, cover crops, like legumes or grass, are planted. Moreover, cover crops increase soil fertility by incorporating organic matter into the soil.

MULCHING: To prevent soil erosion, a layer of organic or inorganic material, such as plastic or straw, is spread over the soil. Mulching also aids in weed suppression and soil moisture retention.

CONTOUR BUFFER STRIPS: To lessen soil erosion and water runoff, vegetation is planted in contour buffer strips along the land's edge. In addition to improving water quality, buffer strips offer habitat for wildlife.

Fertility Management of Soils

Sustaining soil fertility is necessary to enable sustainable farming. Techniques for managing soil fertility include:

CROP ROTATION: This practice includes planting a variety of crops on the same plot of land in a predetermined order. Crop rotation increases crop yields, lowers insect and disease populations, and improves soil fertility.

ORGANIC MATTER ADDITION: Enhancing soil structure, water retention, and nutrient cycling can be achieved by adding organic matter, such as compost or manure. Additionally, organic matter feeds soil organisms, which are vital to the health of the soil.

NUTRIENT MANAGEMENT: Using fertilizers and other appropriate nutrient management techniques helps guarantee that crops have enough nutrients to flourish. However, overuse of fertilizers can cause water contamination and nutrient runoff.

Improving Soil Health

Enhancing soil health is necessary for agriculture to be sustainable. Techniques for enhancing soil health include:

REDUCED TILLAGE: During planting and cultivation, as little soil disturbance as possible is done. Soil organic matter content, water infiltration, and soil structure are all enhanced by less tillage.

CROP DIVERSIFICATION: This refers to the practice of cultivating multiple crops in one field. Crop variety lessens the burden of pests and diseases while enhancing nutrient cycling, both of which contribute to improved soil health.

INTEGRATED PEST MANAGEMENT (IPM): IPM is a pest and disease management strategy that combines chemical, biological, and cultural control techniques. IPM lessens the need for pesticides and lessens the damage they do to the health of the soil.

Protecting the environment and promoting sustainable agriculture depend on maintaining the soil. Farmers may preserve soil health and productivity for future generations by putting into practice soil conservation strategies like fertility management, erosion control, and soil health enhancement. For agricultural areas to be sustainable over the long term and to protect the environment, sustainable soil practices are essential.

Methods of Sustainable Agriculture

A holistic approach to farming, sustainable agriculture seeks to produce food in a way that is socially and environmentally responsible while also being commercially successful. The main goals of sustainable agriculture practices are to protect the environment, reduce the amount of artificial inputs used, and improve the health and welfare of local populations and farmworkers. This chapter will examine some sustainable agriculture practices that support soil protection and health.

ECOLOGY OF AGRICULTURE

The goal of Agroecology is to construct resilient and sustainable agriculture systems by modeling natural ecosystems. Crop diversification, integrated pest management, and agroforestry are examples of agroecological activities. These methods support increased biodiversity on farms, healthier soil, and less reliance on artificial inputs.

CROP DIVERSIFICATION: This refers to the practice of cultivating a range of crops on one land or farm. By lessening the impact of pests and diseases, enhancing nutrient cycling, and encouraging beneficial soil organisms, this technique contributes to increased soil health.

AGROFORESTRY: This technique entails incorporating shrubs and trees into farming systems. Agroforestry may boost biodiversity on farms, give crops and livestock shade and shelter, and improve soil fertility.

INTEGRATED PEST MANAGEMENT (IPM): IPM is a pest and disease management strategy that combines chemical, biological, and cultural control techniques. IPM lessens the need for synthetic pesticides and lessens the environmental damage they cause.

Agriculture for Conservation

The goals of conservation agriculture are to reduce soil disturbance, preserve soil cover, and vary crop rotations. Water conservation, erosion reduction, and improved soil health are all facilitated by conservation agricultural techniques.

Minimal Soil Disturbance: By employing no-till or reduced tillage methods and minimizing tillage, conservation agricultural techniques help to limit soil disturbance. This lessens erosion and enhances water infiltration while preserving the organic content and soil structure.

Preserving Soil Cover: The goal of conservation agriculture techniques is to maintain crop residues or cover crops on the soil all year round. This aids in weed suppression, better soil moisture retention, and erosion protection.

Diversifying Crop cycles: To enhance soil health and lessen the burden of pests and diseases, conservation agriculture encourages the use of various crop cycles. Additionally, diversified crop rotations can enhance nutrient cycling and boost climate change resistance on farms.

Growing Food Organically

Using natural inputs and methods to manage crops and livestock is known as organic farming. Genetically modified organisms (GMOs) and industrial fertilizers and pesticides are not permitted in organic farming. The goals of organic farming are biodiversity, soil health, and environmental sustainability.

Soil Health: By incorporating organic matter into the soil through the use of cover crops, compost, and manure, organic farming techniques contribute to better soil health. This enhances nitrogen cycling, water retention, and soil structure.

Biodiversity: By avoiding the use of synthetic fertilizers and pesticides that may harm soil organisms and beneficial insects, organic farming supports biodiversity. When comparing organic farms to conventional farms, higher levels of biodiversity are frequently observed.

Environmental Sustainability: By avoiding the use of synthetic inputs that can contaminate waterways and harm wildlife, organic farming methods help to lessen the environmental effects of agriculture. Moreover, organic farming contributes to soil carbon sequestration, which may slow down global warming.

Integrated Agriculture

The goal of the permaculture design approach is to replicate natural ecosystems to build self-sufficient and sustainable human dwellings. Using permaculture principles, farming systems can be made more resilient and diversified.

Design Principles: Utilizing renewable resources, watching and engaging with natural systems, and capturing and storing energy are all part of permaculture design. Crop systems that are both productive and sustainable can be developed with the aid of these ideas.

Integration of Elements: To establish a productive and harmonious system, permaculture promotes the integration of many elements on farms, including plants, animals, and structures. This integration contributes to increased biodiversity, decreased pests and illnesses, and improved soil health.

Regenerative Agriculture: Regenerative agriculture, which aims to enhance and restore soil health, water retention, and biodiversity, is frequently linked to permaculture. Because regenerative agricultural techniques store carbon in the soil, they can aid in reducing the effects of climate change.

Utilizing sustainable agricultural practices is crucial for advancing soil conservation and health. Agroecology, conservation agriculture, organic farming, and permaculture are a few techniques that farmers can implement to increase the long-term sustainability of their farming systems. These methods support the health and wellbeing of farmworkers and communities while also lowering the impact on the environment and protecting natural resources. In addition to helping farmers, sustainable agriculture benefits society at large by ensuring a safe and long-lasting food supply for coming generations.

Management of Urban Soils

An essential component of sustainable urban development is urban soil management. Because of pollution, compaction, and impermeable surfaces, urban soils have particular difficulties. Urban soils can be enhanced to enable healthy plant development, lessen the effects of urban heat islands, and increase biodiversity, though, with the right management techniques. Sustainable soil management techniques for urban soils, such as soil testing, remediation, and urban agriculture, will be covered here.

Analysis and Testing of Soils

An important initial step in managing soil in urban areas is soil testing. Tests on the soil can reveal important details regarding its pH, nutritional content, and pollutants. With the use of this information, soil management techniques that are suitable for enhancing soil health and promoting plant growth can be identified.

Nutrient testing: This method assists in identifying the various nutrients present in the soil, such as micronutrients, phosphorus, potassium, and nitrogen. Utilizing this data, fertilization schedules and plant nutrition requirements are established.

pH Testing: This method can be used to assess the soil's acidity or alkalinity. pH testing is a useful tool for determining whether the soil is suited for plant growth, as most plants need a pH range of slightly acidic to neutral.

Contaminant Testing: This method assists in locating any contaminants, such as pesticides, heavy metals, or petroleum hydrocarbons, that may be present in the soil.

Testing for contaminants is necessary to identify any possible threats to the environment and public health.

Methods of Soil Remediation

Pollutants from urban runoff, including pesticides, heavy metals, and petroleum products, frequently contaminate urban soils. To improve the condition of the soil and lessen or eliminate these toxins, soil remediation procedures are applied.

Phytoremediation: This method involves using plants to draw toxins out of the soil. Hyper accumulators are those plants that have the capacity to absorb and retain large concentrations of pollutants in their tissues. Contaminated soils can be cleaned up with these plants.

Bioremediation: This technology breaks down pollutants in the soil by means of microorganisms. Organic pollutants can be broken down by microorganisms like fungi and bacteria, transforming them into less toxic forms.

Soil Amendment: To enhance the structure, fertility, and water-retention capacity of the soil, materials like compost, manure, or bio char are added. Compaction can be lessened and soil health can be enhanced by soil amendment.

Agricultural Practices in Urban Areas

The process of cultivating food in urban environments is known as urban agriculture. Urban agriculture can support regional food production, lower food miles, and increase food security. However, problems with pollution and compaction are common in urban soils, and these can hinder plant growth.

Raised Bed Gardening: This method entails building raised beds and filling them with a composted soil mixture. Raised beds give plants a good growing habitat, aid in better drainage, and lessen compaction.

Container Gardening: Plants are grown in pots, barrels, or buckets as part of container gardening. Growing a range of plants, such as flowers, herbs, and vegetables, in containers is a great way to make use of limited space in urban environments.

Vertical Gardening: Growing plants vertically on walls or other structures is known as vertical gardening. Urban locations with little room can grow plants through the use of vertical gardening, a space-saving strategy.

Ecological Infrastructure

Natural and semi-natural systems that have positive effects on the environment, society, and economy are referred to as "green infrastructure." Green infrastructure techniques can lessen the effects of urban heat islands, manage storm water, and enhance soil health. Examples of these techniques include rain gardens, green roofs, and permeable pavement.

Green Roofs: Made of vegetation, green roofs help to enhance air quality, lessen storm water runoff, and provide habitat for wildlife. In addition to improving building insulation, green roofs lower energy expenses.

Rain Gardens: To assist absorb and filter storm water runoff, rain gardens are small depressions filled with native flora. Enhancing biodiversity, lowering floods, and improving water quality are all made possible by rain gardens.

Permeable Pavement: This kind of pavement permits water to seep through the top layer and into the soil beneath it. Recharging groundwater, lessening heat island effects, and reducing storm water runoff are all made possible by permeable pavement.

In order to improve urban people' quality of life and encourage sustainable urban development, urban soil management is crucial. Cities may enhance soil health, encourage plant growth, and build a more resilient urban environment by putting sustainable soil practices like soil testing, remediation, and urban agriculture into practice. Building sustainable cities for the future requires the use of sustainable soil practices.

Conclusion: A Recap of the Main Ideas

We have examined many facets of soil management, conservation, and sustainable agriculture during our thorough investigation of sustainable soil practices. As we draw to a close, let's review the main ideas we discussed:

1. THE VALUE OF HEALTHY SOILS

Maintaining global food security, sustaining ecosystem services, and promoting plant growth all depend on healthy soil. Microorganisms, insects, and worms are abundant in healthy soils and are essential for maintaining soil structure, cycling nutrients, and managing pest populations.

2. METHODS FOR CONSERVING SOIL:

Many methods of conserving soil have been talked about, such as reduced tillage, terracing, cover crops, and contour plowing. These methods aid in the prevention of soil erosion, the enhancement of water retention, the improvement of soil structure, and the long-term health of the soil.

3. MANAGING SOIL FERTILITY:

Sustaining soil fertility and promoting robust plant growth depend on efficient management of soil fertility. Crop rotation, the addition of organic matter, and nutrient management are examples of practices that help restore soil nutrients, maximize nutrient availability, and reduce environmental effects.

4. SUSTAINABLE FARMING PRACTICES:

The goal of sustainable agriculture practices is to increase agricultural output while reducing adverse environmental effects. In farming systems, practices like permaculture, organic farming, Agroecology, and conservation agriculture enhance soil health, biodiversity, and resilience.

5. MANAGEMENT OF URBAN SOIL:

Sustainable urban development, urban agriculture, and green areas all depend on effective urban soil management. Urban livability is increased, environmental pollution is reduced, and soil quality is improved through the use of techniques including remediation, green infrastructure, and soil testing.

6. ECOLOGICAL INFRASTRUCTURE:

In urban settings, green infrastructure practices—such as permeable pavement, rain gardens, and green roofs—offer advantages for the environment, society, and economy. These methods enhance soil health and biodiversity, lessen the effects of urban heat islands, and control storm water runoff.

7. DIFFICULTIES AND POSSIBILITIES:

Although there are many advantages to sustainable soil techniques, there are drawbacks as well, including low adoption, resource shortages, and legislative obstacles. Governments, academics, farmers, and civil society must work together to support sustainable agricultural initiatives and encourage sustainable soil management techniques in order to address these issues.

8. PROSPECTIVE ROUTES:

Looking ahead, creativity, teamwork, and comprehensive methods are key components of sustainable agriculture and soil management. Innovative approaches to soil conservation, fertility management, and sustainable land use practices can be developed by combining contemporary technologies, conventional wisdom, and multidisciplinary study.

Next Developments in Soil Science

Future research on soil science is expected to be extremely important in tackling issues like food security, environmental sustainability, and climate change. The future of soil science is being shaped by technological advancements, multidisciplinary research, and creative methods, which are opening doors for sustainable soil management and agricultural techniques. We will look at a few of the major themes and advancements that will probably influence soil science in the future in this conclusion.

1. Precision farming and digital soil mapping:

Digital soil mapping (DSM) is transforming soil mapping, analysis, and management. To produce comprehensive soil maps, DSM integrates geospatial technology, remote sensing, and soil survey data. Making educated decisions on soil management techniques, including precision agriculture, can be aided by these maps for farmers and land managers. Precision agriculture optimizes inputs like water, fertilizer, and pesticides based on the unique requirements of each soil and crop by using cutting-edge technologies, such as GPS-guided machinery and sensors.

2. Soil Health Evaluation and Monitoring:

Monitoring and evaluating the health of the soil is becoming more and more crucial for sustainable soil management. Farmers are now able to monitor soil health in real-time and make well-informed decisions about soil management practices because to advancements in soil sensing technologies, such as soil moisture sensors and soil health testing kits. Farmers are also using soil health assessment methods, like the Soil Health Card program in India, to evaluate and gradually improve the quality of their soil.

3. Adaptation and Mitigation of Climate Change:

In order to mitigate and adapt to climate change, soils are essential. By utilizing techniques like cover crops, agroforestry, and no-till farming, soil carbon sequestration can help lower atmospheric carbon dioxide levels and slow down global warming. Adapting to the effects of climate change, such as droughts and floods, can also be facilitated by soil management techniques that increase water retention and decrease erosion.

4. Research on Soil Microbiomes:

Studying the soil microbiome is revealing the intricate relationships that exist between plant health and soil microbes. Scientists are now able to identify and describe soil microbes in unprecedented detail because to developments in DNA sequencing technologies. The function of soil microbiomes in nutrient cycling, plant-microbe interactions, and soil health are being better understood as a result of this research.

5. Management of Urban Soil:

The importance of maintaining soils in urban contexts is growing as the world's population becomes more urbanized. Urban livability is being improved, pollution is being reduced, and soil quality is being improved through the use of urban soil management techniques such brownfield redevelopment, green infrastructure, and urban agriculture.

6. Planning for Sustainable Land Use:

In order to balance conflicting demands for land, such as those from urban expansion, agriculture, and conservation, sustainable land use planning is crucial. Approaches to integrated land use planning, such as ecosystem-based management and landscape planning, can aid in making sure that soils are managed sustainably and that the advantages of soil resources are shared fairly.

7. Governance and Policy:

Policies and governance structures that are conducive to effective soil management are necessary. As the value of soils becomes more widely acknowledged, governments, international organizations, and civil society are creating programs and policies to support sustainable soil management. For instance, the Food and Agriculture Organization (FAO) of the United Nations established the

Global Soil Partnership to advance sustainable soil management and increase public understanding of the significance of soils.

8. Training and Developing Capabilities:

Building capacity and providing education are essential for guaranteeing that the next generation has the information and abilities required to manage soils responsibly. The major providers of instruction and training in soil science, sustainable agriculture, and land management are universities, research centers, and extension agencies.

Glossary

AGROECOLOGY: A farming approach that mimics natural ecosystems to create sustainable and resilient agricultural systems.

AGROFORESTRY: The integration of trees and shrubs into agricultural systems to improve soil fertility, provide shade, and increase biodiversity.

BIOREMEDIATION: The use of microorganisms to break down contaminants in the soil, reducing their harmful effects.

COMPOST: Organic matter that has been decomposed and recycled as a fertilizer and soil amendment.

CONSERVATION AGRICULTURE: A farming approach that aims to minimize soil disturbance, maintain soil cover, and diversify crop rotations to improve soil health and reduce erosion.

COVER CROPPING: The practice of planting cover crops, such as legumes or grasses, to protect and improve soil health between main crops.

DIGITAL SOIL MAPPING: The use of geospatial technologies and soil survey data to create detailed soil maps.

INTEGRATED PEST MANAGEMENT (IPM): A holistic approach to managing pests that combines biological, cultural, and chemical control methods.

NUTRIENT MANAGEMENT: Practices that optimize the use of nutrients in the soil to improve crop yield and reduce environmental impacts.

ORGANIC FARMING: A farming approach that relies on natural inputs and practices to manage crops and livestock, avoiding synthetic pesticides and fertilizers.

PERMACULTURE: A design system that seeks to create sustainable and self-sufficient human habitats by mimicking natural ecosystems.

PHYTOREMEDIATION: The use of plants to remove contaminants from the soil.

SOIL HEALTH: The capacity of soil to function as a living system, supporting plant and animal productivity, biodiversity, and environmental quality.

SOIL MICROBIOME: The community of microorganisms living in the soil, including bacteria, fungi, and archaea.

SOIL PH: A measure of the acidity or alkalinity of the soil, which affects nutrient availability and plant growth.

SOIL TEXTURE: The relative proportions of sand, silt, and clay particles in the soil, which determine soil characteristics such as drainage and fertility.

SUSTAINABLE AGRICULTURE: Farming practices that are environmentally friendly, economically viable, and socially responsible, aiming to preserve natural resources and promote food security.

URBAN AGRICULTURE: The practice of growing food in urban areas, often using innovative techniques such as rooftop gardens and vertical farming.

URBAN SOIL MANAGEMENT: Practices for managing soils in urban environments, including soil testing, remediation, and green infrastructure.

WATER CONSERVATION: Practices that reduce water use and promote efficient water management in agriculture.

www.ingramcontent.com/pod-product-compliance
Lightning Source LLC
Chambersburg PA
CBHW071052240526
45471CB00015B/1649